# Exploring Agriscience

**Delmar Publishers
is proud to
support
FFA Activities**

---

# Exploring Agriscience

## RAY V. HERREN

## Delmar Publishers

I(T)P™ An International Thomson Publishing Company

Albany · Bonn · Boston · Cincinnati · Detroit · London · Madrid
Melbourne · Mexico City · New York · Pacific Grove · Paris · San Francisco
Singapore · Tokyo · Toronto · Washington

# NOTICE TO THE READER

*Cover photos courtesy of:* James Strawser, The University of Georgia (top, middle left, middle right, bottom right); USDA Natural Resource Conservation Service (bottom left); Georgia Pacific (bottom center).

*Cover design courtesy of:* Dartmouth Publishing, Inc.

*Delmar Staff*

Publisher: Tim O'Leary
Acquisitions Editor: Cathy L. Esperti
Senior Project Editor: Andrea Edwards Myers

Production Manager: Wendy A. Troeger
Marketing Manager: Maura Theriault

Copyright ©1997
By Delmar Publishers
a division of International Thomson Publishing Inc.

The ITP logo is a trademark under license

Printed in the United States of America

For more information, contact:

Delmar Publishers
3 Columbia Circle, Box 15015
Albany, New York 12212-5015

International Thomson Publishing Europe
Berkshire House 168-173
High Holborn
London, WC1V7AA
England

Thomas Nelson Australia
102 Dodds Street
South Melbourne, 3205
Victoria, Australia

Nelson Canada
1120 Birchmont Road
Scarborough, Ontario
Canada M1K 5G4

International Thomson Editores
Campos Eliseos 385, Piso 7
Col Polanco
11560 Mexico D F Mexico

International Thomson Publishing Gmbh
Königswinterer Strasse 418
53227 Bonn
Germany

International Thomson Publishing Asia
221 Henderson Road #05-10
Henderson Building
Singapore 0315

International Thomson Publishing - Japan
Hirakawacho Kyowa Building, 3F
2-2-1 Hirakawacho
Chiyoda-ku, 102 Tokyo
Japan

1  2  3  4  5  6  7  8  9  10  XXX  02  01  00  99  98  97  96

**Library of Congress Cataloging-in-Publication Data**

Herren, Ray V.
    Exploring agriscience / Ray V. Herren.
      p.    cm.
    Includes index.
    Summary: Introduces students to the industry of agriculture, covering such topics as plant structures, raising crops, livestock, and poultry, forest science, environmental protection, biotechnology, and more.
    ISBN 0-8273-5809-1 (textbook)
    1. Agriculture—Juvenile literature.   [1. Agriculture.]
I. Title.
S495.H62   1996
338.1'0973—dc20
                          95-18610
                            CIP
                            AC

# Contents

# Preface

*E*xploring Agriscience is intended as the first text that introduces students to the dynamic industry of agriculture. Research has shown that most Americans have a misconception of agriculture. Many think of agriculture as farming. While this is a basic component of agriculture, the scope of agriculture is much more broad. This industry employs more people than any other and is responsible for much of the nation's wealth. This text provides students with an overview of the different aspects of that industry. Of course, all aspects of agriculture are not covered. To address all phases of the industry would take many volumes and would be beyond the confines of classroom teaching. Rather, this text introduces the main areas of agriculture that have made the industry great.

All aspects of the agriculture industry are based on science. Principles of biology have been researched, manipulated, and used to provide advancements in the production of food and fiber. These advancements have made life better for all Americans. No education would be complete without an understanding of the role of agriculture in our lives.

This book was written with the hope that students will continue the study of the science of agriculture. Within each chapter are highlights of particular FFA activities for high school students. These are by no means the only programs of the organization. They represent areas that coincide with the content area of each chapter. Any student with an interest in biological science will find a program in Agriculture Education and the FFA that will match that interest.

This new and exciting title from Delmar Publishers is written to be used as a core text for middle school agriscience curriculum.

# Features

▼ Full color photos, illustrations, and design stimulate student interest and make difficult concepts easier for student comprehension.

**78** CHAPTER 5

FIGURE 5-7
Insects, such as the lady beetle, devour other insects. *Courtesy of USDA ARS K 4249-04.*

FIGURE 5-8 Insects go through a life cycle.

Egg

Adult

Larva

Pupa

Metamorphosis

as the praying mantis and the lady beetle. In some areas these insects are raised for the purpose of eating insects on crops.

Cultural practices have a big impact on diseases and insects. Simply pruning out dead and diseased limbs makes a difference. Burning these diseased limbs and weeding around plants helps a lot. This helps to break up the life cycle of insects. Insects go through several phases in their lives. The cycle begins when the female lays eggs. The eggs hatch into worm-like organisms called larvae. Most of the harm insects do comes about during this stage be-

*The Livestock Industry* **233**

FIGURE 13-12
Annual consumption of lamb is only about 2.5 pounds per capita in the United States. *Courtesy of James Strawser, The University of Georgia.*

. Americans seem to have never or the stronger flavored mutton. of producing lambs for market uality lambs can be produced ing to be fed a lot of expen- increasing number of lambs in feedlots, roughages still t of all the feed consumed

sheep production are , Colorado, South Da- o, Utah, and Oregon. f lower-quality forage they can be success- lands of the desert drier climate helps blems associated ore humid areas. ed according to . Wool type is wool, medium ol breeds are r slaughter. to produce ire, South

FIGURE 13-13
The Dorset sheep is an example of a medium wool breed of sheep. *Courtesy of James Strawser, The University of Georgia.*

▼ Boxed articles entitled "Career Development Events" feature FFA activities and competitions in each chapter.

▼ Numerous hands-on activities and discussion questions reinforce learning of material and concepts covered in the text.

94 CHAPTER 6

be used in flower arrangements. They are cut at the stems so they can fit into vases and other containers. This differentiates cut flowers from flower-

ing plants. Flowering plants may be sold as potted plants. The plants that produce cut flowers are not . Only the flowers and stems are marketed.

Floriculture 95

## Career Development Events

The FFA Floriculture Career Development Event is a competition for teams of four members. Each member competes individually and the final team score is the total of the three highest individual scores. The event consists of four parts: 1) identification of plant materials, 2) placing classes, 3) general knowledge, and 4) practicum.

The identification of plant materials consists of plants selected from the State Floriculture list and is displayed for visual identification. Students may not touch the live plant material but can use a magnifying glass to inspect the foliage. The judging classes are plant materials or floral arrangements. The student places the materials from best to worst based on its structural form, market value, and health. Floriculture plants are selected from the identification list and may be cut flowers, foliage plants, bedding plants, or potted plants. The general knowledge part of the floriculture event consists of 50 multiple-choice

questions. This phase of the event tests knowledge and understanding of all areas of the floriculture industry. Students preparing for this part of the event must

The Floriculture Career Development Event tests students' ability to work with floral arrangements. *Courtesy of Carol Duval, National FFA Organization.*

three individual scores are added to yield the team score. If there is a tie between two team scores, the written exam score is used as a tiebreaker for first and second place teams. State winning teams advance to the National Floriculture Career Development Event.

The FFA Floriculture Proficiency Award is one of the four horticulture proficiency awards. The award is presented to individual FFA members who have excelled in experiences using the principles and practices of field or greenhouse production of flowers, foliage, and related plant materials for ornamental purposes. To apply, the student must be involved in a Supervised Agricultural Experience Program and have maintained a list of floriculture skills he or she has learned. The student must also develop an inventory of buildings, equipment, tools, and materials used in the experience program. He or she must keep accurate records of income, and accomplishments. Students compete within their chapter, state. The state award winner for national recognition.

# Enhanced Content

▼ Provides a hands-on exploratory, science-based approach to agriscience designed to be used as a student's first text in agriculture.

▼ Broad-based coverage of horticulture, animal science, environmental science, and biotechnology is presented in an easy-to-understand format for middle school students.

270 CHAPTER 15

## 15 CHAPTER REVIEW

### Student Learning Activities

1. Make an inventory of all the foods in your house that contains eggs. Do not forget such foods as pastries and mayonnaise. Share your list with the class.

2. Over a period of a week, keep tract of all the meat products eaten by your family. Calculate the percentage of meals that contain poultry products. Include processed meats such as luncheon meats and hot dogs that may contain chicken.

3. Research the possibilities of raising exotic birds such as emus and ostriches. Provide a rationale for growing the birds and the advantages and disadvantages. Report to the class.

### True/False

1. ___ Humans have been growing poultry for almost s long as they have been involved with agriculture.

2. ___ Each year, the average person in the United States consumes almost 90 pounds of chicken and 22 dozen eggs.

3. ___ At one time, most farmers raised chickens for both meat and eggs.

4. ___ Most of the nation's broilers are grown in the northern states.

5. ___ Almost all of the broilers produced are tan.

6. ___ The vast majority of the eggs sold in the United States are white eggs.

7. ___ Turkeys cannot tolerate cold weather as well as chickens.

8. ___ In China and Southeast Asia, ducks and geese make up a large part of the overall diet.

The Poultry Industry 271

Chickens make up the largest part of the poultry industry. The second most widely produced birds are geese.

...ch year thousands of quail and pheasants are released into the wild ...rovide birds for hunting.

### ...hoice

...s considered to be healthier than most meats is

... the poultry industry is raising

...o two parts, broiler production and

... called

... broilers is

# Extensive Teaching/Learning Package

The complete supplement package was developed to achieve two goals:

1. To assist students in learning the essential information needed to continue their exploration into the exciting field of agriscience.

2. To assist instructors in planning and implementing their instructional program for the most efficient use of time and other resources.

## Instructor's Guide to Text

The Instructor's Guide provides answers to the end-of-chapter questions and additional material to assist the instructor in the preparation of lesson plans.

## Lab Manual

### Order #0–8273–6991–3

This comprehensive lab manual reinforces the text content. It is recommended that students complete each lab to confirm understanding of essential science content. Great care has been taken to provide instructors with low cost, strong content labs.

## Lab Manual Instructor's Guide

### Order #0–8273–5810–5

The Instructor's Guide provides answers to lab manual exercises and additional guidance for the instructor.

## Computerized Testbank

**Order #0–8273-7581-6**

An all new computerized testbank (IBM and MAC compatible) with more than 500 new questions that will give the instructor an expanded capability to create tests.

## Teacher's Resource Guide

**Order #0–8273-7494-1**

This new supplement provides the instructor with valuable resources to simplify the planning and implementation of the instructional program. It includes transparency masters, motivational questions and activities, answers to questions in the text, and lesson plans to provide the instructor with a cohesive plan for presenting each topic.

# Acknowledgments

The author gratefully acknowledges many individuals who assisted in the writing of this text. Among them are: Frank Flanders for his advice and assistance; Jean Kilnoski for writing Chapter 17 on companion animals and Chapter 20 on careers in agriscience; Jodi Lange and Lori Nelson for their assistance in writing questions for the end of each chapter; Jim Strawser for his contributions of illustrations; Lynn Barbour for constructing the glossary; Terri Hamlin for writing the inserts on FFA; Wendy Thurston for her contributions in the area of horticulture; Maynard Iverson for his encouragement and support; and a special thanks to Thomas J. Gagliano, Gagliano Graphics, Albuquerque, New Mexico for the illustrations.

I wish to especially acknowledge my wife, Gwen, my son, Bryan, and my daughter, Arenthia. Without their support, this publication would not have been possible.

## Dedication

This book is dedicated to my Aunt Flossie Herren. It was she who gave me my first book and kept me supplied with material to read. She dedicated her life to teaching children and was a constant source of encouragement to me.

# The World of Agriculture

## Student Objectives

**When you have finished studying this chapter, you should be able to:**

- Discuss why agriculture is important in our everyday lives.

- Explain why agriculture is such a broad and diverse industry.

- Explain why agriculture is a science.

- Summarize the reasons why agriculture contributes to the wealth of the nation.

- List and discuss the uses of plants.

- List and discuss the uses of animals.

- Contrast agricultural exports and imports.

- Discuss the processing and distribution of agricultural products.

You are about to begin a study of one of the most dynamic industries the world has ever known! This industry—American **agriculture** —is involved with growing plants and animals for use by all the people of this country and much of the world. We live in a country with one of the highest standards of living in the world. Americans are the envy of

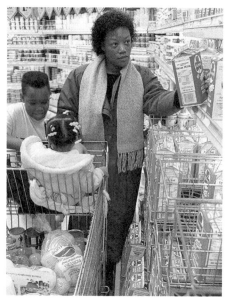

**FIGURE 1-1**
Americans only spend about 17 percent of their income for food. *Courtesy of James Strawser, The University of Georgia.*

most of the world! Much of the affluence we enjoy is either a direct or indirect result of the agriculture industry. Compared to people in other countries we pay a very small portion of our income for food. We spend around 17 percent of our income on food. This compares to over 70 percent in some countries. Our prosperity began because of the fertile soil and ideal climate we enjoy in the United States. Our soil and climate provided the basis for our great agriculture industry. New discoveries and developments as a result of scientific research in agriculture promoted the growth of the agriculture industry. This industry has been one of the major causes of the wealth of the country.

## Science and Agriculture

Science is the study of or the explanation of natural phenomena. In other words, science deals with the laws of nature and our understanding of them. Agriculture fits this definition very well. No other discipline deals more directly with the laws of nature than does agriculture. Recognized areas of science such as physics, chemistry, geology, meteorology, and **biology** all are a part of agriculture. All of the knowledge we have of living things (biology) has only three applications: medicine, **ecology,** and agriculture. In fact both medicine and ecology (the study of the relationship of organisms in an environment) are related to agriculture. Medicine is related to agriculture because good nutrition is dependent on humans obtaining plenty of wholesome food. Ecology is related because of the tremendous impact agriculture has on the environment.

Agriculture is the most important of all the sciences because we depend on agriculture for the food we eat. Without a steady intake of food we would soon die. Through the scientific developments that have made agriculture so efficient, less

FIGURE 1-2
One American farmer feeds over 100 people. *Courtesy of James Strawser, The University of Georgia.*

than half of the people in this country are involved with getting food to the rest of the population. Our system has become so efficient that one American farmer feeds over 100 people. Of course, a large number of people support the farmer in providing supplies and distributing the produce. This allows all of the people not involved in agriculture to produce manufactured goods and conduct services that would not otherwise be possible. If it were not for our agricultural system people would spend most of their time trying to gather something to eat.

Many people think of agriculture in terms of a farmer out plowing a field, and this is a part of agriculture. However, farmers represent a small percentage of the people who work in agriculture. The agriculture industry hires more people each year than any other industry. These jobs involve such diverse occupations as marketing specialists, chemical manufacturers, food inspectors, and research scientists. Yet all of these people are a part of agriculture.

The growing of plants and animals involves many different segments. At one time the growing of living things was only for food and clothing. In modern times in this country many other uses are made of the plants and animals we grow. In fact

FIGURE 1-3
The industry of agriculture hires more people than any other industry. *Courtesy of James Strawser, The University of Georgia.*

new uses are constantly being found. For example, some scientists think that in the future we will be able to grow plants that will supply the fuel needed to run our automobiles!

## Growing Plants

In a very real sense, the foundation of all agriculture is the growing of plants. Without plants, the animals we grow for food would have nothing to eat. Much of the food we eat comes from plants. The flour that goes into our breads and the oil that we use for cooking both come from plants. The fresh fruits that supply us with vitamins and other nutrients are plant products. Just think of all the vegetables that you ate during the past week! All of these were produced by our agriculture industry. The seeds of plants supply us with a rich source of protein. Did you know that peanut butter is made from the roasted and ground up seeds of a plant?

Plants also supply us with clothing. The fiber from cotton plants gives us comfortable clothing. This crop has been grown in this country since before it was a country. Today, only China grows more cotton than the United States.

**FIGURE 1-4**
Agriculture supplies us with nutritious food. *Courtesy of James Strawser, The University of Georgia.*

**FIGURE 1-5**
The growing of trees is a part of agriculture. *Courtesy of James Strawser, The University of Georgia.*

Our houses also come from plants. Trees supply the lumber for buildings and furniture. You might not have thought of forestry as being agriculture, but most of the trees that are harvested each year have been planted and cared for by tree producers. Trees also supplied the paper to make this book. Just think of all the paper Americans use every day. This industry represents a large portion of the economy of the nation.

Many medicines come from plants. A good example is the drug digitalis that is used to help people with heart problems. This important medicine comes from the foxglove plant, which is grown by agricultural producers.

One large and expanding aspect of agriculture is the growing of ornamental plants. Americans spend millions of dollars each year buying plants to beautify their homes. Drive through almost any neighborhood and you will see a broad variety of plants used to enhance the value of homes. Trees, shrubs, grass seed, and turf are all grown on farms and nurseries. Potted plants and flowers are produced in greenhouses for people to buy. Each year many people make their living raising plants in greenhouses.

**FIGURE 1-6**
Plants like this foxglove plant provide us with medicines. *Courtesy of James Strawser, The University of Georgia.*

Plants are used in ways you probably never even considered. Did you realize that chewing gum comes almost entirely from plants? The gum is made from a substance called chicle that comes from the sap of a tree. The sugar used to sweeten it comes from sugar beets or sugar cane. Even flavorings such

## Career Development Events

**FFA**

When you enroll in high school agriculture education courses, you become a member of the FFA. This organization gives you opportunities you may have never even dreamed possible. The skills and concepts you learn in the classroom and FFA activities are a fun way to learn. You can even travel to other countries to learn about agriculture.

The first FFA International Exchange Program was with England in 1948. Since then the FFA international program has grown along with the realization that a world food supply is an important part of the agriculture in the United States. FFA international activities are designed to link FFA members with other nations to provide experiences that will improve agriculture education around the globe. There are several ways in which an FFA member can become involved in an international exchange program.

In your local FFA school chapter you can be part of the international program by promoting and conducting FFA development projects. The development projects can involve single FFA members and chapters nationwide. The FFA can help by improving agriculture practices and crops in developing countries, helping world hunger problems, establishing rural agriculture education programs, and the development of new technology in agriculture. Projects for FFA chapters range from raising funds to support projects to purchasing books, tools, or seed for a rural agriculture education facility.

The chapter may even sponsor and house an exchange student from another country. The World Agriscience Studies (WASS) is an international agricultural student exchange program. Students from other nations are placed with American host families to study agriculture in the United States. The local as mint come from cultivated plants. All these plants are planted, cultivated, and harvested to produce chewing gum. This makes chewing gum an agricultural product!

The growing of the plants is supported by other parts of agriculture. Fertilizers are produced that

FFA chapter can provide foreign students with experiences in agriculture by involving them in their local supervised agriculture experience projects. FFA members also have the opportunity to travel to a foreign country, live with a host

There are several ways for an FFA member to become involved with international activities. *Courtesy of Carol Duval, National FFA Organization.*

family, and attend a year of high school studying an agricultural curriculum.

Other ways to become involved with international agriculture are through study seminars and work experience programs. Agricultural study seminars are organized annually for state associations. The seminars last two to three weeks and usually include much traveling. The seminars typically include visits and tours through agribusiness industries, research labs, agriculture schools, and a wide diversity of farms. Work Experience Abroad (WEA) is an international experience program for FFA members in college or alumni FFA members who work in the agriculture industry. FFA members are placed in other nations to live and work on farms or agribusiness for three to twelve months. The program also coordinates the hosting of students from other nations who will come to the United States. The experience of hosting an exchange student can be as meaningful as traveling abroad.

FIGURE 1-7
The sugar we use comes from crops such as sugar cane. *Courtesy of John Nasniak, Louisiana State University Agricultural Center.*

supply the plants with nutrients. Chemicals are manufactured to produce pesticides that kill weeds and insects that attack the plants. Tractors, cultivators, and harvesting equipment have to be designed and built. Fuel and lubricants have to be produced for the tractors and equipment. Large irrigation systems have to be built, installed, and maintained. All of these items have to be marketed and sold to the producer. The growing of plants is big business!

FIGURE 1-8
All of the machinery used in agriculture has to be designed and built. *Courtesy of John Deere Company.*

# The Animal Industry

Most of the animals produced in this country are raised for food. The cattle industry supplies us with beef. The steaks you eat for dinner and the hamburgers you buy at the fast food restaurant come from the cattle industry. The swine industry gives us such meat products as pork chops, ham, and sausage. The sheep industry supplies meat for a large part of our population. Fish, shellfish, and other aquatic animals are grown by producers to be used as food. The dairy industry supplies us with fresh milk, cheese, butter, cream, yogurt, ice cream, and other products.

The animals that are slaughtered also give us other products. Leather for making shoes, belts, and other clothing comes from cattle, hogs, and other animals. In addition, several medicines and other by-products are made from animal products.

Animals are still used for work. In many places in the western part of our country horses are used to herd cattle. They also are used as pack animals. Horses provide us with recreation through riding and horse racing. Dogs are also produced for work.

**FIGURE 1-9**
Steak for our dinners comes from the cattle industry. *Courtesy of Progressive Farmer.*

**FIGURE 1-10**
Horses provide us with recreation. *Courtesy of Michael Dzaman.*

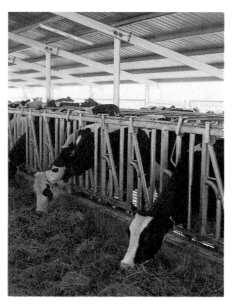

**FIGURE 1-11**
The feed industry provides feed for livestock. *Courtesy of James Strawser, The University of Georgia.*

They are valuable in herding sheep and are also used in moving cattle and hogs.

An increasing part of the animal industry is the raising of companion animals. Not only are dogs raised and trained to be used to help visually impaired people but are raised for pets. The pet industry is very broad and diverse. Animals from goldfish to cats are raised for companions to humans.

Like the plant industry, the animal industry has a lot of support areas. Animals have to be fed. The feed industry supplies producers with high-quality feeds that contain the nutrients animals need. Medicines are also needed to keep the animals healthy. Pesticides are needed to keep the animals free from parasites. Fencing has to be produced. Buildings and facilities have to be built to house the animals. Trailers and trucks are needed to transport the animals. Buyers, brokers, inspectors, and processors are all involved in the industry. And it is all a part of agriculture.

## Agricultural Exports and Imports

An **export** is a product that is shipped from the United States to a foreign country. An **import** is a product that is brought into this country from another country. Agriculture in this country both exports and imports **commodities.** Commodities are products that have value and can be transported. The trade balance of agricultural products is good news for our country. The trade balance refers to the value of goods shipped out of this country (exports) compared to the value of goods shipped into this country (imports). In recent years some industries have lost a large share of their market to companies in other countries. However, American agriculture still dominates the world's agricultural

**FIGURE 1-12**
Agriculture in this country imports and exports goods. *Courtesy of James Strawser, The University of Georgia.*

demand. Our wheat, corn, soybeans, and other products are shipped all over the world to feed people.

Some agricultural products are brought into this country. Although we have a near ideal climate for agricultural production, some crops grow better elsewhere. For example, coffee grows best in the mountainous areas of South America. Bananas grow in the tropical areas of Central America, and many specialty foods such as cocoa and vanilla are brought in from other countries.

Another reason for importing agricultural products is so that we can enjoy fresh produce all year round. Since much of South America is south of the

**FIGURE 1-13**
Coffee is a crop that does not grow well in the United States. *Courtesy of James Strawser, The University of Georgia.*

equator, their seasons are opposite from ours. This means that when it is winter here it is summer there. We take advantage of this by bringing such produce as grapes into the United States so we can enjoy them all year round. Also, countries such as Mexico that are closer to the equator, have a longer **growing season** and can produce vegetables and melons earlier than producers in the United States. By shipping them in from Mexico, we can enjoy watermelons even before it is time to plant them in most parts of this country.

All of the industries involved in exporting and importing agricultural products create a lot of jobs. This adds to the economy and well-being of our nation. When the imports of manufactured products exceed the exports, agricultural exports help to make up the gap.

## Distribution

If you are an average person you probably eat beef from Iowa, apples from Washington State, cheese from Wisconsin, tomatoes from California, potatoes from Idaho, and oranges from Florida. You wear clothes made of cotton from Arizona and live in a house built from lumber grown in Oregon. Of course, other states also produce these products—agricultural products come from a wide area. The states mentioned are ideal for the production of these items because of their climate and other factors. Cotton won't grow very well in Washington State and Arizona is more suited to growing cotton than timber.

Agricultural products are grown where the climate and other factors are suitable for production. However, people in Arizona need lumber and people in Washington State need cotton. To solve this problem, agriculture has a marvelous system for

**FIGURE 1-14**
Agricultural products are shipped from one part of the country to another. *Courtesy of James Strawser, The University of Georgia.*

distributing products all over the country. Many people make their living buying farm produce from producers and selling to wholesale distributors or to processors. These people process and package the produce and sell to retail stores that offer the produce to the public. All this involves a complicated network of transportation. The produce is carried by railroad or by truck. Produce that is easily spoiled may be transported in a refrigerated railcar or truck.

Food and fiber processing is a huge industry that is a part of agriculture. Consumers want products that are as nearly ready to eat as possible. People no longer want to wash, shell, and cook beans in the hulls as they come from the producer. Instead they prefer to have the beans processed and frozen in packages that go into the microwave oven and come out ready to eat. Few people want to make a pizza from scratch when they can have one that is ready to be popped into the oven for a few minutes. Think a minute about all of the processing that had to go into making that pizza. The tomato sauce had to be made from raw tomatoes and spices. The crust was made from flour that was processed from wheat. The cheese had to be pro-

**FIGURE 1-15**
Americans prefer to buy their food ready or almost ready to eat. *Courtesy of James Strawser, The University of Georgia.*

cessed from milk, and the sausages had to be processed from various meats and spices. All of the processing required that large numbers of people work together just to get the pizza ready for you—and they are all a part of agriculture.

## Summary

Look around you. Agriculture is very much a part of your life. Without it our lives would be dramatically changed. In the following chapters we will explore the wonderful world of American agriculture.

# 1 CHAPTER REVIEW

## Student Learning Activities

1. Make a list of all the jobs in your community that are involved in agriculture. Choose a particular job and list all the qualifications one would need for the job.

2. Go to the library and research the agricultural products grown in your state. Write an essay on these products. Which are the most important? Are the products used within the state or are they shipped to other places?

3. Interview ten people about their perceptions of agriculture. Draw a conclusion as to whether or not the majority of the perceptions are correct.

## True/False

1. ___ Physics, chemistry, geology, meteorology, and biology are all part of agriculture.

2. ___ The foundation of all agriculture is growing plants.

3. ___ Most of the animals produced in this country are raised for producing dairy products.

4. ___ U.S. agricultural imports greatly exceed exports.

5. ___ Compared to people in other countries, we pay a very large portion of our income for food.

6. ___ Soil and climate provide the basis for the agriculture industry.

7. ___ A large portion of the medicines we use comes from plants.

8. ___ No other country produces more agricultural products than the United States.

9. ___ Companion animals assist farmers with work on the farm.

10. ___ One large and expanding aspect of agriculture is the growing of ornamental plants.

# Multiple Choice

1. The study of or the explanation of natural phenomena is called
   a. agriculture.
   b. science.
   c. farming.

2. One American farmer feeds over
   a. 100 people.
   b. 200 people.
   c. 500 people.

3. The country that leads in the production of cotton is
   a. the United States.
   b. India.
   c. China.

4. Most of the animals produced in this country are raised for
   a. work.
   b. dairy products.
   c. food.

5. No other discipline deals more directly with the laws of nature than does
   a. agriculture.
   b. horticulture.
   c. floriculture.

6. Agriculture is the most important of all the sciences because we depend on it for
   a. dairy products.
   b. jobs.
   c. food.

7. The foundation of all agriculture is
   a. raising animals.
   b. growing plants.
   c. growing fruit.

8. The seeds of plants supply us with a rich source of
   a. carbohydrates.
   b. protein.
   c. vitamins.

9. Each year many people make their living raising plants in
   a. greenhouses.
   b. their homes.
   c. gardens.

10. Pesticides are needed to keep animals free from
    a. predators.
    b. parasites.
    c. poor weather conditions.

# Discussion

1. How are medicine and ecology related to agriculture?

2. List some jobs involved with agriculture.

3. Why are some plants so important in our daily lives?

4. What are companion animals?

5. What is needed in order to raise healthy animals?

6. List the uses we have for animals.

7. Why is agriculture the most important of all sciences?

8. Where does chewing gum come from?

9. Why is food and fiber processing such a large part of agriculture?

10. List the three applications available for all of our knowledge about living things.

# The History of Agriculture

*When you have finished studying this chapter, you should be able to:*

- Describe the importance of agriculture.

- Discuss how agriculture was necessary to the development of civilizations.

- Tell how the cotton industry brought wealth and created problems in the United States.

- Describe how the cattle industry developed.

- List some of the inventions that had an impact on agriculture.

- Discuss how scientific research helped develop agriculture.

- Define the Land Grant concept of education.

- Explain how the Land Grant institutions have helped shape agriculture.

## KEY TERMS

civilizations
hunters and gatherers
scientific research
Land Grant Act
USDA
experiment stations

The science of agriculture is almost as old as the human race. In the history of humans on the planet, many **civilizations** have come and gone. Some were powerful and warlike and some were small and peaceful. Some civilizations such as the Greeks and Romans provided the foundation for future civi-

**FIGURE 2-1**
Before any civilization can flourish it must have a strong agriculture. *Courtesy of Darl Snyder, The University of Georgia.*

lizations. Architecture, art, literature, and political theories came from these civilizations. Other peoples accomplished great feats of engineering and building. Many aspects of ancient civilizations were different, but they all had one thing in common. Before any civilization could exist and flourish, a strong agricultural base had to be established. The reason for this was simple. Before roads could be built, buildings designed or works of art created, the people had to be fed.

Prior to the creation of agriculture people had to spend most of their time hunting and gathering food. These people are known as **hunters and gatherers.** In order to survive, they had to find the food nature had to offer and collect it. This was a full-time job for all people and little time could be spent doing anything else. This caused the people to have to travel all the time because food in an area would soon be depleted. When people found they could grow their own food many of their problems were solved. Not only did they have a ready supply of food, but they could now settle down in one place.

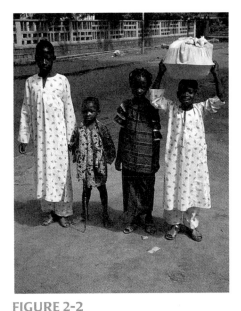

**FIGURE 2-2**
When people could grow their own food they could stop wandering and build villages. *Courtesy of Darl Snyder, The University of Georgia.*

Shortly after people began to plant seed and tame animals, they began to search for better ways of growing food. By trial and error they discovered the best time to plant, the best plants to grow and the best animals to raise. As they got better at growing plants and animals, more food could be produced. As more and more food was produced, it took fewer people to grow the crops and keep the animals. In time, only certain people grew food and, others were then free to accomplish other things. A system of trade was established where goods such as pottery or clothing were traded for the food produced by the farmers. When farmers became really proficient, people could build roads, towns, and create art. This was because other people produced the food needed to feed the workers and builders. The stronger the agriculture of a civilization, the stronger could be the army and groups of workers. Without plenty of food to feed all of the people little progress can be made by any people.

## American Agriculture

The United States is among the wealthiest nations on the earth. Much of this wealth can be attributed to the tremendously successful agriculture system. As pointed out in the first chapter, American agriculture is the envy of the world. No one even comes close to matching the agricultural production of this country.

Our agricultural system began before we were a country. Most of the immigrants who first came to the New World were farmers. The colonists made their living producing crops that could be sent back to Europe. At the time of the American Revolution, well over 90 percent of the colonists made their living through agriculture. At this time the main crops grown were tobacco, sugarcane, and rice. The produce that was not used by the colonists themselves

**FIGURE 2-3**
American agriculture is the envy of the world. *Courtesy of John Deere Company.*

was sent to England, where the products were exchanged for manufactured goods. The rich soil and climate made the New World an ideal place for agriculture. These conditions allowed the development of a variety of crops. But before agriculture could advance very far, certain developments had to take place. The following is a description of some of these developments.

## King Cotton

Perhaps the first huge impact that agriculture had on the wealth of the nation came about as a result of cotton production. Cotton had been grown in India, Egypt, and China for thousands of years. It had also been grown in the New World in what is now known as Peru and Mexico since 2500 B.C. It was introduced into the coastal areas of Georgia and South Carolina in very early colonial days. The fiber from the plant was much prized as a material to use in making clothes. It was easy to spin and weave, easy to dye, and comfortable to wear.

There were two types of cotton—upland and sea island cotton. Sea island cotton was the most widely grown. The fibers were longer (called stable length)

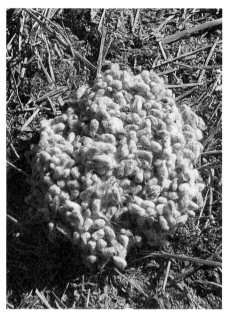

**FIGURE 2-4**
Upland cotton has seeds that seem to be glued to the lint. *Courtesy of James Strawser, The University of Georgia.*

and the seeds could be easily removed from the fibers. The problem with upland cotton was that the fibers seemed to be "glued" to the seeds and were extremely difficult to remove. Even though it was a slow process, the seeds from the sea island cotton separated from the seeds by hand very easily. The problem was that this type cotton would only grow along the coasts of the southern colonies.

About 1790, an inventor named Eli Whitney built a machine called a cotton engine (the name was later shortened to cotton gin) that could remove the seeds from upland cotton. The device consisted of spikes on a revolving drum that passed through slots and separated the lint from the seeds. A rotating brush then removed the lint from the seeds. One person turning the crank of the gin could remove more seeds than many dozen people removing them by hand. When the gin was fully developed and powered by steam, removing seeds was no longer an obstacle to growing cotton. This new device had a profound effect on the history of our country. For the first time upland cotton could be profitably grown. This meant that a large area from what is now southern Missouri to Virginia and

**FIGURE 2-5**
The cotton gin removed seed from the cotton lint. *Courtesy of National Cotton Council of America.*

all of the areas south of there could be used to produce cotton. Cotton was an ideal crop to grow. Only so much tobacco, sugar, and rice could be used and the market could be easily flooded. But cotton was a product that every one could use. Now that the seeds could be efficiently removed, a gigantic market opened for cotton. As the demand grew, more cotton was produced. When this happened, the price fell. This caused the cloth to be affordable to poor people as well as rich. This type clothing was cheaper than the traditional wool or linen. It also had characteristics such as comfort that people wanted. When the price fell more people could buy it and demand increased.

As more and more cotton was produced new fields to the west in Alabama, Tennessee, and Mississippi were cleared and planted. A family could clear the ground, plant, cultivate and harvest enough cotton to make a living. While a lot of cotton was produced by large plantations using slave labor, even more cotton was produced by families

**FIGURE 2-6**
One family could clear the land and make a living producing cotton. *Courtesy of National Cotton Council of America.*

who worked the land themselves. The entire southern portion of the United States was settled primarily because of the cotton industry. Cotton became so important that it became known as King Cotton.

Most of the cotton produced went to Europe. The textile factories of England, France, and

## *Career Development Events*

FFA

The National FFA is an organization for students studying agricultural education in high school. This organization is active in all 50 states and Puerto Rico. Over 450,000 students participate in activities and programs sponsored by the FFA.

This organization began in 1928 when students studying agriculture met in Kansas City for the purpose of establishing a national association. Prior to this, many states had created similar organizations for rural young people. The state organizations began to compete in livestock judging and looked for a place to hold a national competition. The American Royal in Kansas City agreed to host the event. This brought the students together for the first time and they decided there was a need for a national organization. In 1928 the

Future Farmers of America came into existence.

The FFA was patterned after the Future Farmers of Virginia. This state association was developed by Henry Groseclose, Walter Newman, Edmund Magill, and Harry Sanders. They saw the need to give rural youth opportunities to develop social graces and leadership skills. Ceremonies for the FFA were designed after the ceremonies of the Masons, and the awards were patterned after the Boy Scouts.

The FFA has passed many milestones since 1928. In 1965 the FFA merged with the New Farmers of America (NFA). The NFA was an organization for black youths studying agriculture. In 1969 girls were allowed to join the organization. In 1988 the name was changed from the Future Farmers of America to the National FFA Organization.

The FFA still meets in Kansas City for the annual convention. Each year over

Germany ran night and day using cotton from the United States. Ships that delivered the cotton returned with manufactured goods that were sold to the people of the southern states. By the 1850s these industries were almost totally dependent on American cotton.

30,000 young people spend several days in competition, leadership seminars, award programs, and instructional and motivational programs. This is the world's largest youth convention. The FFA is constantly updating and expanding programs to better serve students.

Whether you wish to enter individual competition, team competition, work toward the several degrees of membership, or travel to countries all across the world, there is a program for your needs. There is a place in FFA for you!

The FFA began in 1928 and was organized for students studying agriculture. *Courtesy of the Agricultural Education Program, The University of Georgia.*

FIGURE 2-7 In the mid-1800s most of the cotton produced in the United States went to Europe. *Courtesy of National Cotton Council of America.*

While the south was developing its economy on cotton, the northern states were developing industries. These two diverse sections of the country developed disputes over the role of the federal government, taxation, and about the issue of slavery. Ultimately these differences led to the American Civil War. The development of agriculture had had a profound effect on the young country!

## The Cattle Industry

The cattle industry in this country had a slow beginning. Beef was expensive. In the areas close to population centers, cattle had to be fed expensive grain or had to be fed on pastures that could be put to better use growing crops. In the newly settled areas of the West, cattle could be raised on vast areas of grass. These areas were suited to the growing of cattle but not crops. The only problem was that the cattle grazing areas were far from the centers of population.

After the American Civil War, western cattle producers decided the best way to get their cattle

**FIGURE 2-8**
In the newly settled territories of the West, cattle could graze on the vast areas of grass. *Courtesy of Texas and Southwestern Cattle Raisers Foundation, Ft. Worth, TX.*

to the large cities was to ship cattle on the new railroads that had recently stretched westward. This would provide a way of getting the live cattle close enough to the consumer. This was necessary because the animals had to be slaughtered and consumed within a short period or the meat would spoil. The big problem was that there were only a few places where the railroad ran. Certain areas such as central and western Texas had no rails at all. Towns like Sedalia, Missouri, and Abilene, Kansas, became established as "railheads" where cattle were

**FIGURE 2-9**
Cattle were driven to railheads where they were shipped to the East. *Courtesy of Texas and Southwestern Cattle Raisers Foundation, Ft. Worth, TX.*

brought for shipment. The only way to get the cattle to the railheads was to drive them. The fabled cattle drives of the West were started in this manner. For a period of about 20 years this was the means of getting cattle to market. This allowed cattle producers to make a profit by using the millions of acres of grass land to graze cattle.

After the railroads were finally completed around the turn of the twentieth century, cattle could be delivered live without the long arduous drives. A problem still remained. Even the rail journey took several days. During this time, the cattle lost weight from stress of the long trip. Producers reasoned that more profit could be made if the cattle were slaughtered close to where they were raised and the meat shipped. However, this was impossible because the meat would spoil before it could reach market.

People had known for centuries that if meat was kept cool it could be kept longer without spoiling. The first attempt at cooling meat was the use of ice that had been cut from frozen lakes during the winter and stored in icehouses. The ice blocks were suspended from the ceiling in the meat storage rooms in an effort to keep the meat cool. This effort was not very successful. The ice melted quickly and did not keep the meat cool enough. Then, during the 1880s, mechanical refrigeration was developed. This allowed storage rooms to be kept cool all year long.

A few years later the refrigerated box car was invented. This innovation allowed the transportation of meat anywhere in the country anytime during the year. Now not only could animals be slaughtered any time of the year, but the meat could be stored for a long period of time. This meant that meat could be distributed to every one in the country. This had the added effect of lowering the price of meat. Americans began to enjoy a healthy diet at a relatively low cost.

# Crop Production

Advances were also made in the producing of crops. As people moved westward, vast areas of fertile land were opened for farming. A particularly fertile area was the grassy plains of the Midwest. When settlers first came into the area, they had difficulty plowing the soil. The heavy soil could not be turned with the wooden plows used by the settlers. A heavy cast iron plow would not work because animals could not pull it through the sod. The problem was solved in the 1830s by a man named John Deere. He developed a plow with a cutting edge and share made from steel like that used for saw blades. The strong tempered steel allowed the plow to be thinner and lighter. Wheels were later mounted to make the plow easier to pull by animals. With this invention, the fertile ground could be plowed under and cultivated.

Another problem was that of harvesting crops. For example, wheat was difficult to harvest because the plants had to be cut by hand. They were tied together into bundles called shocks and stacked. They were then gathered and threshed until the grain left the plants. This took a lot of time and reduced the amount of wheat that could be grown.

**FIGURE 2-10**
Settlers had difficulty plowing the thick sod of the Midwest. *Courtesy of Dr. David C. Hartnet.*

**FIGURE 2-11**
The tempered steel of Deere's plow allowed it to be lighter. *Courtesy of John Deere Company.*

**FIGURE 2-12**
McCormick's reaper made harvesting grain more efficient. *Courtesy of Case-International Harvester.*

In 1831, Cyrus McCormick invented a machine that was used for reaping wheat. The horse-drawn implement cut wheat that previously had to be cut by hand. It operated by means of a sickle cutter that was powered by the turning of the wheels. Soon threshing machines were developed to separate the grain from the rest of the plant. This machine was run by steam power and people fed wheat shocks into the machine and it threshed the wheat grain out. After the invention of the internal combustion engine, the process of cutting the wheat and threshing it was generally combined into one operation with a machine that came to be known as a combine.

The development of the internal combustion engine had a revolutionary effect on agriculture. After its development, work that once took days using human or animal power could now be done in a matter of minutes and with a much better result. Today almost all operations involved in the production of agricultural products are mechanized.

## Scientific Research

All of the developments mentioned so far had a real impact on American agriculture. However, by far the most important aspect of the development

FIGURE 2-13
A modern combine cuts and threshes the grain. *Courtesy of John Deere Company.*

was scientific research. **Scientific research** is the systematic search for new knowledge. As mentioned earlier, better ways of growing plants and animals were developed by trial and error. However, this method was too costly and took too long. Dramatic progress was not made until the development of methods of scientific research.

Systematic scientific research began in this country about the middle of the 1800s. It had become apparent that agriculture was a major economic basis for the nation. Leaders thought that our abilities should be strengthened and the best way to do that was to educate people about agriculture. At that time the universities in the United States taught a curriculum known as the classics. Students studied such subjects as Latin, Greek, history, philosophy, and mathematics. Although these were challenging disciplines, they had little practical use. People began to realize that there was a need for institutions of higher learning where students could study subjects such as agriculture that had a practical application.

In the late 1850s a senator from Vermont, Justin Morrill, introduced a bill to provide public land and funds to establish universities to teach practical methods of producing food and fiber. The bill

**FIGURE 2-14**
In 1862 Land Grant universities were established. *Courtesy of Agricultural Education Department, The University of Georgia.*

passed in 1862 and became known as The **Land Grant Act** or the Morrill Act. This act provided each state with public land to build a "college for the common people." The purpose of these colleges was to teach agriculture and mechanics. Many of these institutions were called "A&M" universities. Many, such as Texas A&M, still use this name. During that same year President Lincoln signed into law a bill that established the United States Department of Agriculture **(USDA)**. Soon almost all of the states in the country established land grant colleges.

As students enrolled and classes began, a severe problem was recognized. People began to realize that there was very little knowledge about agriculture. Most of the knowledge about growing plants and animals had been obtained through trial and error. Most of the ideas about agricultural production had no real scientific basis. To solve this problem, Congress in 1872 passed the Hatch Act, which authorized the establishment of **experiment stations** in states with land grant colleges. The purpose was to create new knowledge through a systematic process of scientific investigation. These experiment stations put to use what has come to be known as the scientific method of investigation. Here a systematic, scientific investigation about how plants and animals grow was conducted. Most of the progress of American agriculture came about as a result of the research at Land Grant universities and their experimental stations.

In 1890, another Land Grant Act was passed that provided additional colleges. These served black people who lived in the South. These colleges and universities have made a tremendous contribution to the research and teaching of agriculture. Today the 1890 Land Grant colleges are open to all people.

In 1914, Congress passed the Smith Lever Act, which set up the Cooperative Extension Service. This legislation placed an extension agent in each county in each state. The purpose was to teach the new information learned from research to people. This completed a system known as the Land Grant concept. This concept was that the purpose of a Land Grant university was to teach, conduct research, and carry the new information to the people in the state through the Extension Service.

In 1917, the Smith Hughes Act established Vocational Agriculture in the public high schools as a means of teaching new methods of agriculture. This act established the class you are now in.

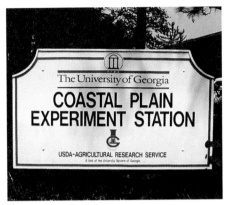

**FIGURE 2-15**
Experiment stations provided research on new and better ways of production. *Courtesy of Stan Diffie, The University of Georgia.*

# Summary

The development of new methods through research and the teaching of this knowledge helped bring about progress in agriculture. Such developments as pesticides, new and better varieties of plants, fertilizers, and soil conservation all resulted from scientific research. As you study the remaining chapters, keep in mind that the agricultural industry is the backbone of our nation. This powerful industry was built as a result of scientific research and development.

# 2 CHAPTER REVIEW

## Student Learning Activities

1. Interview several elderly people who have lived in your community for many years. Ask them about the importance of agriculture to the development of the community. Report your findings to the class.

2. Pick a particular segment of agriculture and do further research on how that segment developed. Discuss your findings with the class.

3. Contact the Cooperative Extension Agent in your area. Discuss some of the latest agricultural research being done or that is recently completed at the Land Grant university in your state. Discuss the impact this research might have on agriculture in your area. Report to the class.

## True/False

1. ___ Before any civilization could exist and flourish, a strong agricultural base had to be established.

2. ___ At the time of the American Revolution, 50 percent of the colonists made their living through agriculture.

3. ___ The main crops during the American Revolution were tobacco, sugar cane, and rice.

4. ___ The rich soil and climate made the New World an ideal place for agriculture.

5. ___ Sea island cotton fibers were extremely difficult to separate from the seeds.

6. ___ Eli Whitney developed a machine to remove seeds from cotton.

7. ___ The northern states developed their economy around industry.

8. ___ President Lincoln signed into law a bill that established the United States Department of Agriculture.

9. ___ Most of the immigrants who first came to the New World were industrialists.

10. ___ Most of the cotton produced went to Europe.

# Multiple Choice

1. The civilizations that provided foundation for future civilizations were the Greeks and the
   a. Romans.
   b. Americans.
   c. Europeans.

2. Crops were sent to England and exchanged for
   a. wheat reapers.
   b. cotton gins.
   c. manufactured goods.

3. In colonial times, the most widely grown cotton was
   a. King cotton.
   b. manufactured cotton.
   c. sea island cotton.

4. The invention that allowed the transportation of meat anywhere in the country, anytime during the year, was the refrigerated
   a. cotton gin.
   b. box car.
   c. "railhead."

5. A machine was developed for reaping wheat by
   a. Cyrus McCormick.
   b. Eli Whitney.
   c. John Deere.

6. The most important aspect in the development of agriculture was
   a. wheat reaping.
   b. scientific research.
   c. colony industrialization.

7. Another name for the Land Grant Act was the
    a. Morrill Act.
    b. Hatch Act.
    c. Hughes Act.

8. The Smith Hughes Act established
    a. manufactured goods.
    b. scientific research.
    c. vocational agriculture.

9. The biggest problem cattle producers faced was
    a. research.
    b. transportation.
    c. climate.

10. The Smith Lever Act set up the
    a. Cooperative Extension Service.
    b. refrigerated box car.
    c. Smith Hughes Act.

# Discussion

1. What problem did cattle producers have?

2. Why is John Deere important?

3. What is the Land Grant Act?

4. What is the Land Grant Concept?

5. Why was cotton so important in the South?

6. Why is agriculture important?

7. Why was agriculture necessary to the development of civilizations?

8. How does the cotton gin operate?

9. How did scientific research help develop agriculture?

10. Why did cotton become so important?

# 3

## Soil: The Origin of Life

### KEY TERMS

soil

food chain

humus

organic matter

minerals

glaciers

erosion

flood plain

texture

soil horizon

ecosystem

## Student Objectives

**When you have finished studying this chapter, you should be able to:**

- Explain why agriculture and all of life depends on the soil.
- Explain the difference between organic and inorganic soil.
- Discuss the carbon cycle.
- Discuss the different ways soils are formed.
- Contrast the differences between acidity and alkalinity and the effect of each on the soil.
- Explain the concept of a soil ecosystem.
- Conceptualize the balance of organisms in the soil.

The earth has a thin coating of material that is called **soil.** Imagine that the earth is an apple. The layer of soil would be as thin as the peel of the apple. Yet without this layer, there could be little life on the earth. Most plant life is dependent on the soil. Plants are rooted in the soil and get their life-sustaining nutrients from it. If

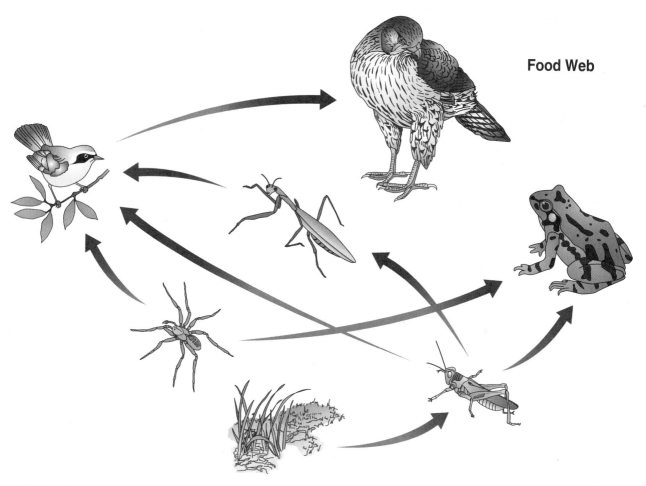

**Food Web**

FIGURE 3-1 In the food chain nutrients are passed from the soil to living creatures and back to the soil.

there were no soil, there would be no plants for land animals to eat. Life is in a constant cycle with the earth. A process called the **food chain** begins with the soil. A plant obtains nutrients from the soil and uses them through energy from the sun. To obtain the nutrients it needs, a mouse may eat the plants. An eagle then eats the mouse and obtains nourishment. Eventually the eagle dies and its body is decomposed. The decaying of the eagle returns nutrients to the soil and the chain begins again. In this process all life is dependent on the soil.

Soil is composed of four ingredients: minerals, air, water, and **humus.** Humus is another word for **organic matter** and **minerals** are substances that did not come from living organisms. Minerals have the chemical and physical properties that can be formed into rocks. Minerals in the soil are very small particles. As these fine particles are grouped together, they form pores or gaps where the particles touch. These pores are filled with either air or water.

# Organic Soils

Organic matter comes from living sources. In the example of the eagle that dies, the material left when it decays is organic material or humus. Soil can be grouped into two very broad categories based on how much mineral or humus the soil contains. Soil that has a lot of humus or organic material is called organic soil. This soil is the result of a buildup of plant materials that grew in the soil, died, and decayed. Over many hundreds of years, the decayed plants turned into a black fertile soil called organic soil. Organic soils exist in parts of the Midwest where centuries of grasses grew and

**FIGURE 3-2**
Organic soil was formed by hundreds of years of decaying grasses. *Courtesy of John Deere Company.*

returned to the soil when they decomposed. Another area with these soils is southern Florida, where swamp land has been drained. For thousands of years, a type of grass called saw grass grew in the water. When it matured, it fell into the water and decayed. Over time, several feet of this organic material turned to soil.

Organic soil is very rich in nutrients. Also, the organic material in the soil allows nutrients in fertilizer to be available for plant use. Organic soils are some of the world's most productive soils.

## Inorganic Soils

Soils that originated from minerals are the result of the wearing down or breaking down of rock. This process is often the result of weather. As wind blows against a rock, particles are loosened and removed. Likewise rain, sleet, and snow on rocks can remove particles that become soil. The process of wetting and drying as well as the dissolving of minerals by the water can also help in the process of breaking down the stone. The freezing and thawing of water in cracks and crevices of rocks can wear away tiny

**FIGURE 3-3**
Wind is a powerful force in weathering rocks and the formation of soil. *Courtesy of Richard Frear, National Park Service.*

particles and can open other cracks for the entrance of more water. The process of weathering of rocks is an extremely slow process that occurs over thousands and even millions of years.

Ice may form soil by moving across the surface of the soil. Thousands of years ago, great rivers of slowly moving ice called **glaciers** extended across the northern portions of America, Europe, and Asia. As these huge masses of ice crept southward, they carried along rocks that were in the path. As the glaciers moved along, the rocks were crushed together with enough force that some were pulverized and made into particles small enough to become soils. With the change of the seasons part of the ice would melt and then freeze again. During warm periods the front of the ice mass would melt and drop all the debris it had picked up along the way. When the weather once again turned cold, the ice resumed its movement and picked up other rocks and soil particles. This mass of large boulders, smaller rocks, sand, silt, and clay is called till. The land mass built up by the dropping of till by a glacier is called a moraine. Obviously the type of debris picked up by the glacier determined the type of soil that was deposited. Moraines vary in the type

**FIGURE 3-5**
Water that moves in a stream often picks up and transports soil.
*Courtesy of USDA Natural Resource Conservation Service.*

and texture of the soil. Some contain a lot of sand and are coarser while some contain the finer silt.

## Water-Deposited Soil

As mentioned earlier, moving water wears away rock and forms soil particles. Soil that has been transported and deposited by moving water is called alluvial soil. Some of the world's most productive soils are the result of alluvial action. Water is a major way that soil is moved from one place and deposited in another. Water in a stream picks up soil from the ground. When rains wash soil particles into the stream, the currents carry the material down the stream. This process is called **erosion.** As long as the water is moving rapidly, the soil particles remain suspended in the water. As soon as a streambed begins to level out and the water slows down, the soil particles settle to the bottom. If the stream flows out of its bank, deposits are left in the area covered by the water. These areas are called **flood plains.** Also, as the stream nears its mouth, the water slows down and the deposits settle to the bottom. These areas are called deltas. Gradually over the years, the soil is built up to a great depth by both of these processes.

In this country, the Mississippi River has carried deposits south for thousands of years and deposited them at the mouth of the river. The result has been the rich farmland of the Mississippi Delta. Since the early history of our country, this area has produced rich crops. It should be noted, however, that the soils that made up the delta were once in the Midwest. The soils in the areas where the soil left are less productive because of the erosion.

Soil may also be deposited into a lake. As a stream empties into a lake, the soil deposits are dropped and they sink to the bottom. Eventually over thousands of years, the lake becomes filled

**FIGURE 3-6**
The rich farmland of the Mississippi Delta is the result of soil transported by water. *Courtesy of John Nasniak, Louisiana State University Agricultural Center.*

with soil deposits. When the lake disappears, a deep fertile soil is left in place of the water.

## Soil Deposited by Wind

The wind also creates and transports soil. In many areas of the world, tall outcroppings of rock have been carved by the wind. As the wind wears away the stone, the loose particles become soil and form deposits known as eolian soils. If the particles are large particles the soil is called sand and the deposits are called dunes. In parts of the western United States, vast areas are covered in huge sand dunes that are constantly moving and changing shape. Since the sand is constantly moving, little vegetation can grow on the dunes.

Fine soil particles like silt and clay that have been deposited by the wind creates soil known as loess soils. These soils may be found through the Mississippi Valley, parts of the Midwest, Washington, and Idaho. Many of these soils were formed when the glaciers melted and the climatic conditions turned dry. When the soil dried out the fine particles were blown by the wind and accumulated in new areas.

**FIGURE 3-7**
Sand dunes constantly move and little vegetation can grow on them. *Courtesy of James Strawser, The University of Georgia.*

In many of the mountainous regions of the world, soil was (and still is) formed by the action of volcanoes. Volcanoes bring deposits from deep within the earth and spew the material on the surface. Much of this material is in the form of molten

## Career Development Events

**FFA**

The FFA land judging Career Development Event is a competition for teams of four members. Each member competes individually and the final team score consists of the total of the three highest individual scores. The FFA land judging contest aids classroom instruction by teaching the students practical application of the Soil Conservation Service classification system. The competition and hands-on applications motivate students to learn soil and water management practices. A student's decision-making skills are developed and environmental awareness increased through the analysis of soil characteristics and the identification of important soil and water resources.

This event tests the FFA members' ability to determine a site's topsoil texture, thickness, and effective depth of topsoil and subsoil. The student must also decide the permeability of subsoil, land slope, erosion, drainage and any other factors that will best decide the land's capability. After the student determines land class, the proper land use and treatment practices must be decided. The student judges the land by the observation and measurement of prepared soil pits and the land area surrounding each pit. Each team member should come to the event equipped with a soil probe, pencil, clipboard, and calculator. After all team members have completed each phase of the event, each individual score will be calculated. State winning teams advance to the National Land Judging Career Event.

These activities help make classes more interesting and give students an additional reason for doing their best to develop necessary skills. The land judging Career Event requires students to study classroom material and practice the hands-on activities. The extra time in preparation for the event not only helps produce a winning team but also helps prepare for a career.

The Soil and Water Management Proficiency offers individual FFA memrock (called lava). This inorganic material flows out over the surface and cools into solid rock. Other deposits are spewed high into the air. The finer of these particles is called volcanic ash and may be carried many miles by the wind. Near the volcano

bers an opportunity to be awarded for experiences related to renewable natural resources. Students who have participated in the planning and use manage-

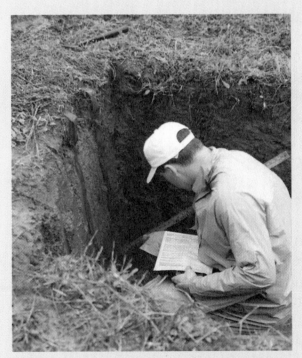

In the Land Judging Career Development Event, students judge the land by the observation and measurement of prepared soil pits and the land surrounding the pits. *Courtesy of Frank Flanders, the Agricultural Education Program, The University of Georgia.*

ment practices to improve air, soil, water, timber, fish or wildlife can apply for this award. Programs in which a member is involved in helping to stimulate public awareness or assisting in educating the public concerning pollution problems should be included in the application. Other activities a student might participate in are the development and implementation of management practices that will prevent erosion, improve the productivity of the soil, promote efficient use of water, or reduce air and water pollution. To compete for the proficiency, the student should keep a list of related skills developed as a result of the above activities and training in vocational agriculture. The student should participate in a Supervised Agricultural Experience Program (SAEP) that relates to soil and water management. Proficiency applications are judged on the student's records, activities, and experiences. Keeping accurate records of accomplishments and soil and water management related income and expenses are essential to becoming a winner.

**FIGURE 3-8**
Soil from volcanic ash is usually very fertile. *Courtesy of Fred E. Mang, Jr., National Park Service.*

many feet of volcanic ash may be built up in a matter of days. Areas as far away as a hundred miles may receive measurable deposits of ash. Soils from volcanic ash are generally very productive soils. Volcanic soils can be found in the states of Hawaii, Oregon, and Washington.

## Soil Texture

The **texture** of the soil refers to the size of the individual soil particles. The larger the size of the soil particles, the coarser the soil feels when rubbed between the fingers. The largest of the soil particles is sand. Sand particles are small but they are considered large compared to other types of particles. If a soil containing sand is rubbed between your fingers, it will feel coarse and gritty, like corn meal. Silt particles are smaller than sand and feel smoother to the touch. The smallest of the particles is referred to as clay. A soil high in clay content will feel smooth, like flour. One way to illustrate the relative sizes is to imagine that if a clay particle was the size of a BB, a silt particle would be the size of a golf ball and a sand particle would be the size of a basketball.

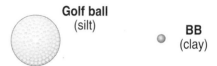

**Relative Sizes of Soil Particles**

Basketball
(sand)

Golf ball
(silt)

BB
(clay)

*Sizes may not be to proper proportion.*

**FIGURE 3-9** If a clay particle were the size of a BB, a silt particle would be the size of a golf ball, and a sand particle would be the size of a basketball.

Soils are seldom composed of pure sand, pure silt, or pure clay. More often they are a combination of the three. Soil that contains less than 52 percent sand, 28 to 50 percent silt and 7 to 27 percent clay is called loam soil. This soil is considered to be near ideal for growing most crops. Soils that have a large concentration of sand are coarse and do not hold water very well. Because the particles are so large, the water from a rain or irrigation system passes on through the soil. Water that goes on through the soil is of little use to plants. Clays and silts slow the water down as it is absorbed into the ground and hold a portion of it that the plants can use. If the soil has a large percentage of clay, the ground will not let enough water through. This may either cause the water to run off or may hold the water too long and cause problems for the plant.

Clay also plays an important role in soil fertility. Most producers apply fertilizers to the soil. Fertilizers provide the nutrients needed for the plant to grow. Clay particles in the soil hold the nutrients in place until the plants can use them.

**FIGURE 3-10**
Soil with a high clay content does not allow water to pass through very fast. *Courtesy of USDA Natural Resource Conservation Service.*

**pH Value**

**FIGURE 3-11**
The pH scale measures the acidity or alkalinity of a substance.

# Soil pH

The pH of the soil is a measure of how acidic it is. Acids contain hydrogen that forms particles called ions when dissolved in water. Alkaline is the opposite of acid. Vinegar is a common substance that is acidic. Baking soda is alkaline. The pH scale goes from 1 to 14. A score of 7 means that the substance is neutral. The lower the number, the more acid the substance is. The higher the number, the more alkaline the substance.

Certain crops may have problems growing in soils that are too acid or too alkaline. Producers frequently adjust their soils to the right pH. They have samples of their soil tested for pH. If the soil is too acid, the correct amount of an alkaline substance (such as lime) is added. If the soil is too alkaline, acid materials (usually sulfur-based materials) are added to the soil. Crops such as blueberries grow well in soils that are relatively acid (pH of 4.0 to 5.0). Crops such as asparagus do well in soils that are less acid (pH of 6.0 to 8.0).

# Soil Horizons

When soil is formed from rock or from organic material the soil forms in layers. These layers lie parallel to the surface of the earth. As layers are added they be begin to take on different characteristics. These different regions are called **soil horizons.** If a pit is dug deeply into the ground, the vertical bank on the side of the hole gives a picture of the different layers or horizons of soil. This picture is called the soil profile.

The uppermost layer is called the O horizon. This layer contains undecomposed and decomposed organic matter. This is the place where most seeds germinate.

The next layer is the A horizon and is composed of the top soil that contains organic matter along with minerals. It is generally darker than the horizons below and may be from a few inches to several feet thick.

The region under the A horizon is the E horizon and is the area of the topsoil that has little organic material. This horizon is light in color because the clays and humus have been washed out by water that passes through the soil.

The next region is the B horizon. In this subsoil region, materials washed out (or "leached") from the other horizons accumulate and form. This area is generally high in clay content. The B horizon combined with the O, A, and E horizons make up a region of the soil where plant roots grow.

Underneath the B horizon is the C horizon that is made up of parent material. Parent material is the material from which the soil originated. This horizon has not been affected by the soil-making

**FIGURE 3-12**
A pit dug into the ground can give a view of the soil profile. *Courtesy of James Strawser, The University of Georgia.*

**FIGURE 3-13** Soil is classified in soil series. Some examples are: (A) Entisols. (B) Aridisols. (C) Iceptisols. (D) Aflisols. *Courtesy of USDA Natural Resource Conservation Service.*

process and does not have the properties of the soil in the other horizons.

The last horizon in the soil profile is the R horizon. This horizon is composed of the bedrock upon which the other soil horizons rest.

## The Soil Ecosystem

Your first observation of the soil might lead you to believe that it is lifeless. Looking across a freshly plowed field, you might think that all that is there are minerals and decaying plant matter. Actually, the soil abounds with life. The soil is an **ecosystem.** An ecosystem is all of the plant and animal life that live in an area. The life-forms depend on each other for the proper balance of food and other environmental factors. Within this ecosystem are many different forms of plant, animal, and microbial life. Most of these forms of life depend on each other for their existence or for a maintenance of the proper balance in nature.

### Plant Life

Most plants depend on the soil for their existence. The soil provides the support for the root system that anchors the plant and makes it stable. Most of the nutrients used by the plant are obtained from the soil. Even though we are used to seeing the top portions of plants, a very large portion of the plant lives in the soil. Roots of most plants can reach as far as several feet into the soil. In fact, such common crops as alfalfa have been known to grow as far as 25 feet or more into the ground. The area of the soil that contains the roots of plants is called the rhizosphere. In this zone of the soil the plant receives water and the nutrients it needs to live and grow.

**FIGURE 3-14**
A large part of a plant lives in the soil. *Courtesy of USDA Natural Resource Conservation Service.*

## Microorganisms

Within the rhizosphere live billions of microorganisms of many different types. Many of these organisms live off the roots of plants. As plants live and grow, they are constantly oozing materials out through the roots. This material contains protein and other materials that the microorganisms feed on. In addition, as root cells mature and die, the microorganisms act to decompose the cells. When the entire plant dies they help to return the nutrient material the plant took from the soil back to the soil. This is known as the carbon cycle.

The most common soil microorganisms are bacteria. They are so abundant that in only one teaspoon of soil there can be as many as 500 million bacteria. Several types of these bacteria live in what is known as a symbiotic relationship with plants. This means that organisms of different types live together for mutual benefit. The most common symbiotic bacteria are the nitrogen fixing bacteria, rhizobia, that live in the roots of certain plants. Plants that host these organisms are called legumes and include such plants as beans, clovers, peanuts, alfalfa, and peas. Rhizobia live in lumps on the roots called nodules. The bacteria live in the soil and attach themselves to the root hairs of the legume shortly after the plant sprouts. As a reaction to the bacteria attaching itself to the root, a gall or knot forms on the plant root. It is within this swelling that the rhizobia bacteria grow and reproduce. They receive all the nutrients they need to live and reproduce from the host plant. In return the bacteria convert nitrogen from the air in the soil into a form of nitrogen that the plant can use.

Fungi are plantlike organisms that contain no chlorophyll. They range in size from microscopic to the large mushroom fungi that grow on the surface of the soil or on decaying plant material. These

**FIGURE 3-15**
Bacteria, called rhizobia, form nodules on roots. They convert nitrogen from the air to a form useful to the plant. *Courtesy of USDA ARS.*

Soil fungi help to decompose plant material. *Courtesy of James Strawser, The University of Georgia.*

organisms are as abundant as bacteria in the soil. They play an important role in the breakdown and decay of plant materials. Fungi are particularly important in forest soils because they break down lignin, which is a primary component of wood. Trees that die and fall to the ground are returned to the soil through the action of fungi, bacteria, and other organisms.

Protozoa are one-celled organisms that live in moist soil. They are aquatic organisms because they live in particles of water in the soil. When the soil becomes dry, the protozoa change into an inactive state until the soil becomes moist again. Protozoa feed on bacteria in the soil. Through this process a balance of bacterial life in the soil is better maintained.

One of the most important group of microscopic animals that inhabit the soil is nematodes. Nematodes are worms in the class Nematoda that have smooth round bodies that are not segmented. Although many nematodes are not microscopic, most of those that live in the soil are so small they can't be seen with the naked eye. These tiny worms are very abundant. In a typical spade full of moist soil there may be over a million nematodes. In fact, they are the most abundant multicelled animals in the soil. There are three basic groups of soil nematodes. One group consumes decaying organic mat-

Nematodes can feed on the roots of plants. They are the most numerous multicelled animals in the soil. *Courtesy of Ed Brown, The University of Georgia.*

ter. Another group eats other microorganisms. The third group feeds on plant roots. The third group is by far the most important because of the damage they do to the plants.

## Summary

Soil is essential to all life. All of our food sources depend on it. Humans must protect it because it takes hundreds or even thousands of years to build soil. The fertile soil of this nation has provided much of our wealth. If we take care of our soil, our wealth will continue.

## 3 CHAPTER REVIEW

### Student Learning Activities

1.  Bring in a sample of soil from around your home. Determine if the soil is organic or inorganic. Also determine the soil texture and if the soil is mostly sand, clay, silt, or loam.

2.  Observe the area around your school. Attempt to determine how the soil might have been formed in the area. What indications did you use?

3.  Make a list of all the ways soil is important to you. Compare your list to others in the class.

### True/False

1.  ___ Most plant life is dependent on the soil.

2.  ___ If there were no soil, there would be no plants for land animals to eat.

3.  ___ A process called the food chain begins with the soil.

4. ___ Minerals have the chemicals and physical properties that can be formed into soil.

5. ___ Organic soil is rich in nutrients.

6. ___ The largest of the soil products is sand.

7. ___ If the soil is too alkaline, acid materials are added to the soil.

8. ___ The most common soil microorganisms are the bacteria.

9. ___ Soil is made up of only minerals and decaying plant matter.

10. ___ Roots of most plants cannot reach very far into the soil.

# Multiple Choice

1. A plant obtains nutrients from the soil and uses them through energy from
   a. water.
   b. food.
   c. sunshine.

2. Humus is another word for
   a. organic matter.
   b. bacteria.
   c. decaying plants.

3. Soil that has been transported and deposited by moving water is called
   a. fertile soil.
   b. alluvial soil.
   c. eolian soil.

4. Fine soil particles like silt and clay that have been deposited by the wind create soil known as
   a. eolian soil.
   b. alluvial soil.
   c. loess soil.

5. In many of the mountainous regions of the world, soil was formed by the action of
   a. volcanoes.
   b. earthquakes.
   c. neither a nor b.

6. Fertilizers provide plants with
    a. insect repellent.
    b. nutrients.
    c. food.

7. Plants that host rhizobia are called
    a. legumes.
    b. fungi.
    c. nematodes.

8. One-celled organisms that live in moist soil are called
    a. protozoa.
    b. nematodes.
    c. fungi.

9. Plant-like organisms that contain no chlorophyll are called
    a. protozoa.
    b. legumes.
    c. fungi.

10. The area of the soil that contains the roots of plants is called the
    a. subsoil.
    b. rhizosphere.
    c. C horizon.

# Discussion

1. Soil is composed of what materials?

2. Why is plant life dependent on soil?

3. How is soil classified?

4. List two ways soil can be deposited.

5. What is loam soil?

6. What is the problem with too much clay in soil?

7. What are soil horizons?

8. What is the carbon cycle?

9. What is a symbiotic relationship?

10. What are the three basic groups of nematodes?

# 4

# Plant Structures and Their Uses

## Student Objectives

*When you have finished studying this chapter, you should be able to:*

- Describe the functions of plant stems.
- Tell the functions of xylem and phloem.
- Explain how stems are used by people.
- Describe the functions of a leaf.
- Explain how leaves are used by humans.
- Discuss the function of flowers.
- Explain how flowers are used by people.
- Explain how bees are used in agriculture.
- Describe the functions of seeds.
- Explain how seeds are used by people.
- Discuss the functions of roots.
- Tell how roots are used by people.

### KEY TERMS

stems
photosynthesis
flowers
nectar
seeds
roots

All life as we know it on Earth is dependent on plants. Without plants, the atmosphere of our planet would not have the oxygen required to support animal life. As well as providing air to breath, plants are essential for food, fiber, erosion control and many other

FIGURE 4-1
Plants provide food for humans and feed for animals raised for food. *Courtesy of James Strawser, The University of Georgia.*

necessities. Almost all of agriculture is dependent on plants. Plants provide food for humans and the agricultural animals raised for food. Plants, like animals, have different parts that serve different functions. People make use of these plant parts in a variety of ways.

Plants are made up of stems, leaves, flowers, seeds, and roots. Each of these parts plays an important role in the life of the plant. Agricultural producers grow particular plants in order to obtain plant parts or structures that can be used.

## Stems

**Stems** are very important to plants. They support the plant and provide a means of moving nutrients from one end of the plant to the other. Some stems grow into large trunks, as in trees. Other stems may grow in the form of twining vines that allow the plant to climb fences or other plants.

At the very tip of the stem is a small knob called the terminal tip containing the bud that will become next season's growth. Further down the stem there are small scars that go all the way around the stem. These mark the place on the stem where last year's terminal tip or bud was. These scars are called bud scale scars. It is easy to calculate how old a stem is by counting these scars.

Every place along the stem where leaves are or were have scars, too. These are called leaf scars and they occur at nodes. Also at the nodes are small buds called axillary buds. These buds may at some time grow into stems or side branches. They grow in the axil formed by the leaf and stem.

Inside the stem are tube-like structures called the xylem and the phloem. The xylem carries the water and nutrients absorbed by the roots to the leaves. The phloem carries sugars manufactured in the leaves down to the roots. Agriculture makes use of these parts. In trees, the woody part is the xylem. Lumber cut from trees is composed of the xylem of

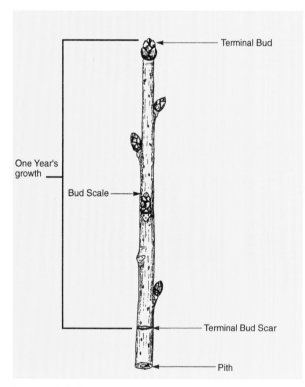

**FIGURE 4-2**
The age of a stem may be calculated by counting the bud scale scars.

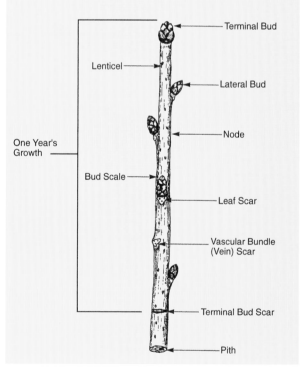

**FIGURE 4-3**
Parts of a stem.

Phloem/inner bark

Xylem/sapwood

Outer bark

Cambium

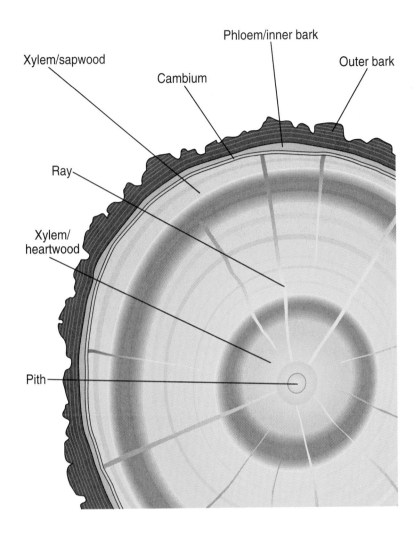

Ray

Xylem/
heartwood

Pith

FIGURE 4-4
The xylem of trees is used to make paper.

the stems. Power poles, fence poles, and plywood is composed of stem materials. Also, the xylem of stems from trees provides fiber for making paper. Just think of all the uses for paper and paper products.

In a plant called flax, fibers are removed and woven into a cloth called linen. These fibers are the phloem of the stems of the flax plant. The outer layers of woody tissue (the xylem) are removed and the phloem is harvested.

Stems are also used extensively for food. The celery we eat in salads or as an ingredient in cooked foods is a stem. Rhubarb and asparagus are also

FIGURE 4-5

Stems of plants like asparagus are used for food. *Courtesy of James Strawser, The University of Georgia.*

FIGURE 4-6

Cattle eat plant stems in the form of silage. *Courtesy of James Strawser, The University of Georgia.*

examples of stems that are used for human food. Potatoes are also a type of stem. These stems, called modified stems, grow underground. They provide a place for the plant to store nutrients. These nutrients, stored in the form of starches, have fed people for thousands of years.

Stems can also be used as animal feed. Plants such as alfalfa are ground up and fed to cattle. Although not as nutritious as the leaves, stems provide a food source. Corn stalks and grain sorghum stalks (along with the rest of the plant) are chopped and made into a feed called silage. Many cattle are fed on silage.

## Leaves

Everyone is familiar with leaves. They provide the beautiful green colors of spring and the reds, yellows, and browns of autumn. Leaves add color and decoration both inside and outside our homes. Potted plants with unusual leaf colors are used inside the home for decoration. Plants that turn different shades and different colors with the seasons are used outside.

Leaves also come in a wide assortment of shapes and sizes. The leaves of a juniper are less than ¼ inch long, but the leaves of elephant ear plants can be as large as 3 feet across. Leaves grow in alternate, opposite, or whorled patterns. In the alternate arrangement, leaves appear on one side of the stem at a node and from the other side at the next node. Opposite leaves grow from opposite sides of the stem directly across from each other. Whorled leaves refer to three or more leaves that grow around the stem at the same node.

Leaves are where the manufacture of food takes place. In a process known as **photosynthesis,** the energy of sunlight is used to produce nutrients. These nutrients are used by the plant for growth, maintenance, and reproduction. This food is in the form of sugars and carbohydrates as well as fats and proteins. Since this food is stored in leaves, animals use the leaves for food. Over the years, agriculturists have discovered and developed plants with leaves that are high in nutritive value and that live-

**FIGURE 4-7**
Some plants like the elephant ear have huge leaves.

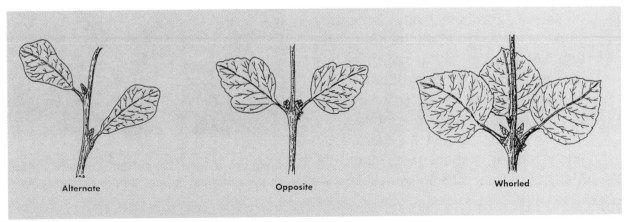

Alternate   Opposite   Whorled

**FIGURE 4-8** Leaves of different plants are arranged differently on the stem.

stock will eat. Cattle, sheep, and horses eat the leaves of grass and other plants as forage. Producers cut and dry these forages into hay that is stored and used as livestock feed.

Humans also eat the leaves of plants. As with plants used to feed animals, scientists have developed those plants that make good food for humans. Cabbage, mustard, lettuce, and spinach are all examples of plant leaves eaten by humans. The leaves are a primary place for food storage in the plant. The food stored by the plant can be used as food for animals. Many of the nutrients needed by animals and humans are obtained from the eating of plant leaves.

## Career Development Events

FFA

The Food for America program is an excellent way to expose your FFA to elementary school children. This program is designed to make young students aware of the products and opportunities of agriculture. Many elementary students have never had the opportunity to visit any agriculture industries, businesses, or farms. The agricultural source of many everyday products, from milk to shoe polish, is easily overlooked.

Few Americans have an accurate understanding of modern agriculture. Many of today's consumers believe that

FFA Members help elementary students understand where food comes from. *Courtesy of Blane Marable, Morgan Co. High School, Madison, GA.*

FIGURE 4-9
Humans also make use of plant leaves for food. *Courtesy of James Strawser, The University of Georgia.*

since the number of farmers is decreasing that agriculture is dying. What they fail to realize is that agriculture technology has helped farmers become more productive. Today's agricultural producer uses less land to provide for more people. The Food for America program exposes the network of the many industries associated with agriculture from production, transportation, and storage to the marketing of food and other agricultural products. Careers as scientists, marketing specialists, engineers, veterinarians, nutritionists, and other occupations are important parts of American agribusiness and should be discussed with elementary students. Through the Food for America program, FFA members help elementary students understand these crucial roles and how food and other agricultural products are produced. FFA members should help elementary school teachers in agricultural instruction and bring examples of animals and agriculture products into the classroom. FFA members may even offer tours of a working farm as a field trip for the elementary students. FFA members are excellent role models. With well-practiced presentations, FFA members can leave a positive image of today's agriculture and the FFA through the Food for America program.

# Flowers

**Flowers** are the more attractive part of the plant. Most flowers are flashy, like the rose or daisy. Others are so plain and simple it is hard to realize they are flowers. Flowers serve as the reproductive organs for many plants. Some plants reproduce asexually. This means that these plants reproduce without a sperm and egg uniting. Usually these type plants do not have flowers. Plants that reproduce sexually have flowers.

Some species of plants have male flowers and female flowers on one plant. Others have female flowers on one plant and male flowers on another plant. Most, however, have both the male and the female parts in the same flower.

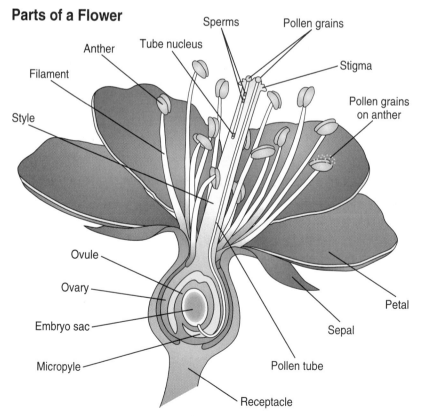

**Parts of a Flower**

FIGURE 4-10
The parts of a flower.

Most flowers consist of the same basic parts. The female part, called the pistil, is made of the stigma, style, and ovary. The male part is called the stamen and contains the anther and filament. The pistil produces eggs that are fertilized by the pollen produced in the stamen. Pollen is usually taken from the anther to the stigma by insects, the wind, or in some cases, animals.

In order to attract the bees or other insects, plants have developed brilliantly colored flower petals. Although insects probably don't see the colors of the flowers the same way as we do, they are attracted by the petals. The flowers secrete a sweet fluid known as **nectar** that the insects use for food. As they crawl or fly into a flower to collect nectar, grains of pollen are caught on their legs or other body parts. When they go to the next flower, the grains of pollen are knocked off and remain in the flower. This causes pollination to occur.

Honeybees are particularly adept at pollinating. Many insects work flowers, but most go from one kind of flower to different kinds of flowers each day. Bees, on the other hand, work a particular kind of flower for a period of time. For example, honeybees will work apple blossoms for several days until the flowers are gone then work a different type of flower. By doing this they go from an apple blossom to another apple blossom and spread pollen from one apple blossom to another. This process ensures that the blossoms are thoroughly pollinated.

Fruit growers hire beekeepers to bring in truckloads of bees in the spring as the trees are blooming. The bees live in wooden, box-like structures called hives with a separate colony of bees in each hive. The hives are easy to handle and can be loaded on a truck with the bees still in the hive. The owner of the bees can then move the hives from orchard to orchard for a fee from the fruit or crop producer. In addition, the producer can harvest

**FIGURE 4-11**
Brightly colored flowers attract insects that pollinate the plants. *Courtesy of James Strawser, The University of Georgia.*

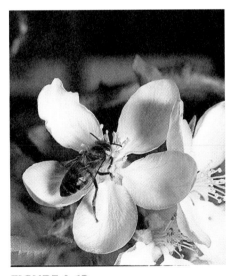

**FIGURE 4-12**
Honeybees are particularly adept at pollinating. *Courtesy of James Strawser, The University of Georgia.*

**FIGURE 4-13**
Commercial honey producers place bees in an orchard to pollinate the trees and to produce honey. *Courtesy of Keith Delaplane, The University of Georgia.*

hundreds of pounds of honey each year that can be sold at a profit.

## Seeds

**Seeds** generally result from the flowers on plants. Seeds are another part of the reproductive process of the plant. Most plants begin their growth from seeds. The production of seed is one of the most important links in the entire agricultural industry. In fact, a lot of our food is derived from seed. Estimates are that around 70 percent of our diet comes directly from the seeds of plants. The bread or bowl of cereal that you had for breakfast this morning was made from the seed of grain plants. Cooking oil is derived from the oil that is pressed from such seeds as safflower, soybean, or cotton seed. Snack foods such as peanuts and popcorn are all made from the seeds of plants.

Seeds are also used for a variety of other purposes. Many of our beverages, such as coffee and cocoa, come from the seeds of trees and shrubs. Oils and dryers in paints and wood finishes come from seeds. Oil from the seed of the castor bean was once used for medicine and is now used as an oil in hydraulic

**FIGURE 4-14**
Breakfast cereals are made from plant seeds. *Courtesy of James Strawser, The University of Georgia.*

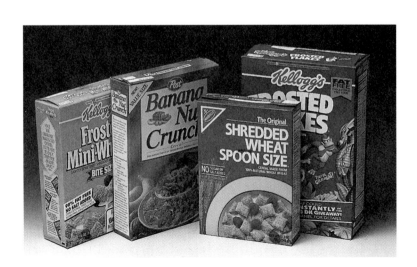

systems. Seeds are also used in the manufacture of cosmetics, ointments, and pharmaceuticals.

Seeds produced on plants often are encased in a fruit. For example, apple seeds develop in the pulpy sweet material that is eaten by people and animals. Many different fruits encase seed. Plums, peaches, pears, and oranges all have seeds at the center. Some foods we don't generally think of as fruits also fit this description. Tomatoes, squash, eggplant, and watermelons all encase seed and are eaten by people.

**FIGURE 4-15**
The fruit of an apple encases seeds.
*Courtesy of James Strawser, The University of Georgia.*

## Roots

The two main functions of **roots** are to support the plant and to take in water and nutrients from the soil. Plant roots are at least as extensive as its stems and leaves. When you see a large tree just think of how much of it is really in the ground, hidden from your view.

There are four main types of roots: fibrous, tap, adventitious, and aerial. Fibrous root systems have many tiny roots that lead from the plant deep into the ground. One corn plant can have as much as 100 miles of roots.

Most trees have tap roots. These are very large tough roots that reach deep into the earth for moisture and keep the tree from falling over. Carrots are also examples of a tap root system. There are many smaller roots that come off the main root.

Adventitious roots usually occur on plants that also have fibrous or tap root systems. An example are corn plants that have roots coming out of the main stalk above the ground that help to brace the plant against strong winds.

Aerial roots are common in very wet areas of the world. These roots are able to obtain water directly from the air. Plants with this type of root system

**FIGURE 4-16**
Humans also make use of roots, such as sweet potatoes, for food. *Courtesy of James Strawser, The University of Georgia.*

don't even need soil to grow in. Both orchids and staghorn ferns have aerial roots.

As well as being a necessary part of the plant, roots are important to humans. Plants are often used to slow down or stop erosion. Roots help to hold soil in place. All of us have eaten roots at one time or other. Carrots, turnips, radishes, and sweet potatoes are all examples of roots that are commonly eaten. Some spices also come from roots. Ginger is a root that is grown in India and Southeast Asia. It has been used for hundreds of years to add spice to foods.

## Summary

Plants have always been used by people to eat, build shelter, and make clothing. Almost all of agriculture is involved with plants. Without plants and their various uses, people would have a difficult time existing. Agriculture has improved plants that were once found in the wild. Chosen for a particular use, these plants were bred to be more efficient. Modern agriculture will continue to produce these plants to fill the needs of people.

# 4 CHAPTER REVIEW

## Student Learning Activities

1.  Make a list of all the food in your house that comes from plants. Determine the part of the plant from which the food was made. Share the results with the class.

2. Look around your home and list all the plants that are ornamental. What parts of the plants provide the beauty?

3. Choose a particular plant part (such as leaves, stems, roots, or flowers) and create a list of all the different uses that could be made of this part.

# True/False

1. ___ Potatoes are a type of stem.

2. ___ Plants reproduce asexually.

3. ___ Leaves serve as the reproductive organs for many plants.

4. ___ Some species of plants have male flowers and females flowers on one plant.

5. ___ Nectar is a pesticide.

6. ___ Aerial roots are common in cold areas.

7. ___ Plants are often used to slow down or stop erosion.

8. ___ Whorled leaves refer to two or more leaves that grow around the stem at the same node.

9. ___ Leaves are where the manufacture of food takes place.

10. ___ In a process know as photosynthesis, the energy of sunlight is used to produce nutrients.

# Multiple Choice

1. At the very tip of the stem is a small knob called the
    a. terminal tip.
    b. node.
    c. xylem.

2. The structure inside the stem that carries the water and nutrients from the roots to the leaves is the
    a. phloem.
    b. node.
    c. xylem.

**3.** Flowers are really just specialized
    a. leaves.
    b. plants.
    c. vegetation.

**4.** Those plants that reproduce sexually have
    a. leaves.
    b. flowers.
    c. stems.

**5.** Honey bees are particularly adept at
    a. asexual reproduction.
    b. producing nectar.
    c. pollinating.

**6.** Most tree roots are
    a. fibrous.
    b. tap.
    c. aerial.

**7.** The xylem of tree stems provides fiber for making
    a. cotton.
    b. linen.
    c. paper.

**8.** The male part of the plant is called the
    a. pistil.
    b. stamen.
    c. stigma.

**9.** Insects are attracted to the
    a. petals.
    b. leaves.
    c. color.

**10.** Seeds produced on plants often are encased in
    a. plants.
    b. soil.
    c. fruit.

# Discussion

1. Why are plants so essential to us?

2. What are the major parts of plants?

3. What are the functions of the xylem and phloem?

4. What are some uses humans make of plant stems?

5. What are the parts of the flower?

6. Why are insects important to flowers?

7. List the main types of roots.

8. Why is the production of seed one of the most important links in the agriculture industry?

9. For what purposes can leaves be used?

10. How are bees used in agriculture?

# Agricultural Pests

## Student Objectives

*When you have finished studying this chapter, you should be able to:*

- Explain why producers have to deal with pests.

- Discuss how insects are harmful.

- Define the life cycle of insects.

- Discuss how insects are controlled.

- Analyze the concept of integrated pest management.

- Explain what a weed is.

- Tell how weeds are controlled.

- Discuss the newest methods used in pest control.

Agriculture deals with the growing and cultivating of living things for the use of humans. People provide the best care possible for the plants and animals they grow. One of the basic parts of caring for crops and animals is protecting them from other living things. In the wild, plants and animals are constantly in competition with each other. In nature all living things need space and nourishment. Animals eat other animals, animals eat plants, and plants get nutrients from the soil. When humans grow plants or animals, they provide an enticing meal for animals such as insects. Every year cropland is prepared for planting. The soil is tilled, fertilizer is spread, and conditions are made favorable for the crop to grow. Unfortunately, the conditions also favor the growth of unwanted plants.

Many industries have developed to protect plants and animals from pests, such as the manufacture of cultivation equipment and the development of agricultural chemicals. These industries use a tremendous amount of resources each year. Machines and chemicals have to not only be made, but have to be distributed to the producer.

**FIGURE 5-1**
A vast industry has developed to protect plants from pests. *Courtesy of James Strawser, The University of Georgia.*

## Insects

**Insects** are what most people think of when they think of agricultural **pests.** Since humans first began growing their own food, insects have been a problem. Many ancient documents tell of troubles with insects, such as locusts, that devoured crops. For thousands of years it was accepted that insects would get a large percentage of the crops. Problems

**FIGURE 5-2**
In the past it was accepted that insects would get a large percentage of the crop. *Courtesy of James Strawser, The University of Georgia.*

FIGURE 5-3
Consumers are reluctant to buy damaged fruit. *Courtesy of James Strawser, The University of Georgia.*

FIGURE 5-4
Insects such as mosquitoes carry disease. *Courtesy of USDA ARS K 4705-01.*

with insects continue even to this day. If insects are not controlled on crops, millions of dollars' worth of damage will be done. Not only will insects eat a large part of the produce, but they will also damage most of the crop not eaten. We are accustomed to seeing crisp fresh apples at the grocery store. These apples are almost flawless and have no damage. However, if insects are not controlled as the apples are growing, the apples we buy will have defects such as worm holes. Most consumers would not buy this type of apple. The same is true of almost all that we grow.

Not only do insects eat our crops, they also carry **disease.** Millions of people have contracted diseases because of insects. Insects such as the mosquito may draw blood from a person infected with a disease and then inject disease organisms in the next person it bites. If the same mosquito bites many people, the disease becomes widespread. Yellow fever and malaria are diseases spread by mosquitoes. Also throughout history countless millions of head of livestock have died as a result of diseases carried by insects.

Some insects are **parasites.** Parasites are organisms that live on other organisms (called hosts) and harm them. Agricultural animals are harmed by such insect parasites as ticks, fleas, and heel flies. These insects get their nutrients from the animal they live on or in. This causes the animal to have to eat more in order to live and do well. In turn, the producer must give more feed to the animal or control the parasite.

One of the biggest problems with insects is that there are so many of them. Over 750,000 species have been classified. Scientists think there may be as many as 2,000,000 more that have not been identified. Of the identified insects, only about 500 are considered to be major pests in this country. However, these pests are formidable enemies. One

**FIGURE 5-5**
One female insect may lay millions of eggs in her lifetime. *Courtesy of James Strawser, The University of Georgia.*

female insect may lay millions of eggs during her life. If most of these hatch out, there will be a lot of insects around to cause trouble. Fortunately, nature helps out. Many birds and animals eat large amounts of insects. Other insects such as the lady beetle and the praying mantis devour huge numbers of insects. Each year vast numbers of insects are eaten by predators.

Even with all the predators eating insects, that still leaves a lot to destroy crops. For many years we have used chemicals to get rid of insects. These chemicals, called **insecticides,** have been very efficient. They have been the most efficient means ever devised to kill insects. In the past insecticides have been used in tremendous quantities. The overuse of insecticides has resulted in problems. Many of the older chemicals like DDT have been harmful to the environment, and wildlife has suffered. These chemicals remained in the soil for many years. Rain would wash the chemicals into streams where they were absorbed by aquatic plants and animals. Animals such as water birds and eagles ate the fish and absorbed the chemicals. Reproductive cycles were disrupted because the eggs from the birds would not hatch.

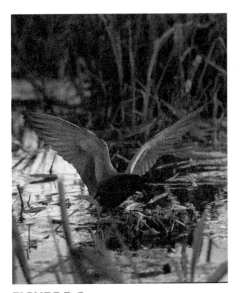

**FIGURE 5-6**
Reproductive cycles of animals such as water birds and eagles were disrupted by DDT. *Courtesy of University of Minnesota Agricultural Experiment Station.*

Today, all insecticides are regulated very closely by governmental agencies. Most of the old insecticides that remained in the soil for so long have been banned. New insecticides break down into harmless substances in a short time. This makes environmental problems less likely because the residues of the chemicals do not build up.

Any sound insect control program involves

## Career Development Events

Agriculture students should be aware of the damage that insects, diseases, and weeds create in all areas of agriculture.

The combination of classroom study and FFA activities prepares students to handle pest problems correctly and safely. Agriculture teachers educate students on methods of identification, control, and prevention of these pests. The FFA offers students the opportunity to compete and be awarded for their knowledge on this topic.

The National FFA Foundation provides funds for recognition of members' achievements at the local, state, regional, and national levels. Awards to members are generally presented at the annual chapter banquet. Competition includes team events and individual efforts to identify plant pests and disorders in forestry, floriculture and nursery/landscape career development events. When students judge plants, they need to be able to look for symptoms of insect, parasite, and disease damage and presence.

FFA members may use a pest-related topic such as "Integrated Pest Control in Agriculture" to compete for the FFA public speaking career development events. Speaking out on agriculture issues not only develops skills in communication but informs the public about changing times. There are two FFA public speaking contests—a six to eight minute prepared speech and an extemporaneous speech.

FFA members have the opportunity to research and write prepared speeches. The speech manuscript is submitted to a panel of judges and used as part of the judging. The extemporaneous speech competition is designed to test the FFA

**integrated pest management** (IPM). This approach combines knowledge of the pest and its life cycle, monitoring, cultural practices, biological control, and chemical pesticides.

Much has been done recently in the area of **biological control.** Biological control involves the use of predatory insects, fungi, or viruses. You are probably familiar with the appetite of such insects

member's ability to talk about a given subject without having prepared or rehearsed its content. The FFA member is given 30 minutes to generate a speech on a given agriculture topic. Both events judge the student's power of expression, stage presence, voice control, knowledge, and response to questions.

In several FFA Career Development Events students must inspect plants for insect damage.
*Courtesy of Carol Duval, National FFA Organization.*

**FIGURE 5-7**
Insects, such as the lady beetle, devour other insects. *Courtesy of USDA ARS K 4249-04.*

as the praying mantis and the lady beetle. In some areas these insects are raised for the purpose of eating insects on crops.

Cultural practices have a big impact on diseases and insects. Simply pruning out dead and diseased limbs makes a difference. Burning these diseased limbs and weeding around plants helps a lot. This helps to break up the life cycle of insects. Insects go through several phases in their lives. The cycle begins when the female lays eggs. The eggs hatch into worm-like organisms called larvae. Most of the harm insects do comes about during this stage be-

**FIGURE 5-8** Insects go through a life cycle.

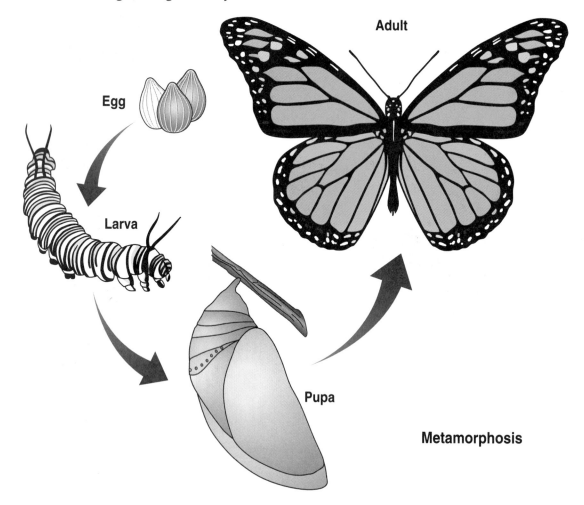

Egg

Larva

Pupa

Adult

Metamorphosis

cause the larvae eat a tremendous amount of plant material. Larvae develop into the pupa stage and may be covered for a period by a cocoon. The pupa then develops into the adult stage. The adult lays eggs and the cycle begins all over again. Depending on the type of insect, they are more vulnerable at one phase than others. For example if they spend the winter as a pupa in crop residue or in dead fruit tree limbs, getting rid of the crop residue or burning the dead limbs will get rid of the insects.

In integrated pest management, a combination of methods are used to control the insects. The crop is closely monitored and chemicals are used only when necessary. This not only cuts down on production costs but also helps protect the environment.

## Weeds

**Weeds** are plants that are growing where they are not wanted. If oats are growing in a wheat field the oats may be considered to be weeds. In fact some of our worst weed pests were once considered to be good plants. One of the best examples is Johnsongrass, which was brought in from Africa. In the 1800s it was thought that this grass would be the

**FIGURE 5-9**
In the 1890s, Johnsongrass was thought to be a good pasture grass. Now it is a serious pest. *Courtesy of William K. Vencill, The University of Georgia.*

perfect pasture grass in the South. It is a fast-growing grass that can be used for grazing. However, the grass got into other fields and competed with crops. It has now spread all across the country and is extremely difficult to kill in crops. It outgrows most agricultural plants and robs them of sunlight and nutrients.

Weeds are pests in many different ways. They can harbor insects and diseases. Remember that insects go through several life phases. They may need weeds as a means of getting through one of the phases. Weeds compete with agricultural plants for nutrients and water. Weeds must have nutrients the same as agricultural plants. If a field is full of weeds, the weeds may get as much or more of the soil's nutrients as the good plants. This means there are fewer nutrients for the crop. Also, the weeds may block out the sunlight. Sunlight is needed by the plants to produce sugars in the process of photosynthesis.

Another problem with weeds is that they may lower the quality of the crop. Hay that has weeds in it may be less nutritious or may even contain poisonous weeds. Grain that contains weed seed is not as good for human food or animal feed. A cotton

FIGURE 5-10
Weeds can block out sunlight from reaching other plants. *Courtesy of William K. Vencill, The University of Georgia.*

crop that has weeds will have foreign particles in the lint and will be less valuable.

Most weeds are hardy, produce many seeds, and can completely take over an area very quickly. The fact that weeds are still around indicates that they are hard to control. People have been fighting them for thousands of years and the fight goes on.

People have used many means to fight weeds. Pulling the weeds up, hoeing, and plowing are the oldest means. Also animals have been used to get rid of unwanted plants. The animals are brought in and confined to an area where they eat the weeds. The most efficient way of controlling weeds is the use of a **herbicide,** a substance that kills plants. Chemical herbicides have been widely used since the 1960s and continue to be a very effective means of weed control.

There are basically two types of herbicides. One is the nonselective herbicides. This type kills all plants that it contacts. These are used in places where all of the vegetation is considered to be weeds. For example, nonselective herbicides are used on railroad beds to kill all plants. They can also be used with crops if the herbicide only contacts the weeds and not the crops.

The other type is selective herbicides. These will kill only certain types of plants. For example, a selective herbicide might kill cockleburs that are growing in corn. The herbicide will not affect the corn plants even if it comes in contact with them.

Years of research have gone into developing selective herbicides. A new herbicide must pass several requirements. First, it must not harm the crop it is used on. Second, the chemical has to be safe to use by producers. Producers come in contact with the herbicides as they are applied. If the herbicide is toxic to humans, the applicator might be harmed. Third, it must not be harmful to the environment. We have to protect plants and animals around us,

FIGURE 5-11
Weed plants usually produce a lot of seed. *Courtesy of James Strawser, The University of Georgia.*

FIGURE 5-12
Selective herbicides kill only the weeds. *Courtesy of James Strawser, The University of Georgia.*

**FIGURE 5-13**
Pesticides have to be safe to apply.
*Courtesy of James Strawser, The
University of Georgia.*

so we have to be careful that the herbicide kills only the plants we intend. Fourth, the chemical must be affordable to use. It would do no good to use a chemical that was more expensive to use than the benefit produced. Modern herbicides must fit all of these requirements before they are released for use.

Some of the newer means of controlling weeds include the use of biological agents. This means that organisms such as insects and disease germs can be used. If an insect's favorite food is a particular weed, the insect might be used to control the weed. Of course, researchers have to be careful that the insect does not develop an appetite for crops too. If the insect meets the requirement of only eating the weed pest, it can be raised for the purpose of eating the weed.

Diseases can also be used to fight weeds. Some diseases are particular about the type of plant they attack. Generally, a particular bacteria, virus, or fungus will attack only a certain type of plant. If a disease-causing agent can be found that only attacks a particular weed, the disease might be effective in controlling the weed.

**FIGURE 5-14** Diseases are a serious problem in plants. The plant shown in (A) is healthy; the one in (B) is diseased. *Courtesy of James Strawser, The University of Georgia.*

# Genetic Engineering

One of the most exciting possibilities for controlling agricultural pests is the use of genetic engineering. Genes are the substances in the cells of plants and animals that contain the material (DNA) responsible for passing characteristics from one generation to the next. Scientists now have the ability to take genes from one organism and place them in another. This means that they can alter the characteristics of plants and animals. Plants have been developed that are resistant to insects. This means that insects will not be as likely to feed on the genetically altered plants.

We now have the ability to make plants more tolerant to herbicides. Obviously, the more tolerant a crop plant is to a herbicide, the more likely weeds can be controlled. One of the problems is that the more similar the crop is to the weed, the more difficult it is to use a selective herbicide. By using genetic engineering, we can make the crop more unlike the weed.

## Summary

Humans have always faced pests that threaten agriculture. Although we spend a lot of time and money growing crops and animals, they are still a part of nature. In nature, plants and animals depend on each other. It is a difficult task to keep agricultural plants and animals away from the other organisms in nature. Through study and experimentation, we will better understand agricultural pests and find new and better ways of controlling them.

# 5 CHAPTER REVIEW

## Student Learning Activities

1. Collect ten insects that are harmful and ten that are not harmful. Identify the insects and tell what damage the harmful insects do.

2. Identify a harmful insect that is common to your area. Devise a means of controlling the insect. Make sure your plan uses the integrated pest management concept.

3. Identify the weeds that are common to your area. Collect five and bring samples to class. Discuss with the class how the weeds spread.

## True/False

1. ___ Insects do not carry diseases.

2. ___ Agriculture deals with the growing and cultivating of living things for human use.

3. ___ One of the basic tasks in caring for crops and animals is protecting them from other living things.

4. ___ Cancer and diabetes are diseases spread by mosquitoes.

5. ___ Some insects are parasites.

6. ___ Diseases can also be used to fight weeds.

7. ___ One of the biggest problems about insects is that most are deadly to humans.

8. ___ Insecticides are the most efficient means ever devised to kill insects.

9. ___ Any sound insect control program involves integrated pest management.

10. ___ Weeds are plants that are growing where they are not wanted.

## Multiple Choice

1. The insect cycle begins when the female
   a. dies.
   b. mates.
   c. lays eggs.

2. Insect eggs hatch into worm-like organisms called
   a. larvae.
   b. pupas.
   c. cocoons.

3. A substance that kills plants is called a
   a. pesticide.
   b. herbicide.
   c. insecticide.

4. Organisms that live on organisms and harm them are called
   a. insects.
   b. parasites.
   c. predators.

5. Many people have contracted diseases because of
   a. insects.
   b. plants.
   c. predators.

6. An estimation of the number of major insect pest species in the United States is
   a. 750,000.
   b. 2,000,000.
   c. 500.

7. Two insects that devour huge numbers of insects are
   a. lady beetles and praying mantises.
   b. ticks and fleas.
   c. cockleburs and honey bees.

8. Johnsongrass is considered to be a
   a. landscaping grass.
   b. perfect pasture grass.
   c. weed.

9. Plants have been developed that are resistant to
   a. insects.
   b. pesticides.
   c. animals.

10. Every year cropland is prepared for
    a. animals.
    b. insects.
    c. planting.

## Discussion

1. List and discuss some of the problems caused by insects.

2. Define integrated pest management.

3. What is meant by the biological control of insects?

4. Why are weeds considered pests?

5. What is the difference between a nonselective and a selective herbicide?

6. How can genetic engineering be used for controlling agricultural pests?

7. How do cultural practices have an impact on diseases and insects?

8. What is the life cycle of an insect?

9. How can weeds be controlled?

10. List two types of biological agents.

# Floriculture

KEY TERMS

floriculture
bulbs
florists
potted plant
foliage plants
house plants
cut flowers
photoperiodic

## Student Objectives

*When you have finished studying this chapter, you should be able to:*

- Define floriculture.

- Describe the national aspects of floriculture.

- Describe the international aspects of floriculture.

- Describe the three basic types of living plants florists sell.

- Describe the steps in producing a typical flower crop.

- List and describe the four flower arranging principles.

The word **floriculture** literally means the growing of flowers. Floriculture is a very big business in the United States and the world. Not only are billions of flowers produced each day, but they must be taken to their destination, arranged, stored, and sold. Each part of this chain is encompassed by the floriculture business.

**FIGURE 6-1**
Floriculture is a big agricultural business. *Courtesy of Washington State University.*

# International Aspect

Flower crops have been grown for centuries in Europe. Throughout history flower gardens have produced beauty. Almost everyone enjoyed growing or looking at the flowers. Today, huge flower markets sell flowers by lots of hundreds and thousands. People seem to enjoy flowers as much as ever.

Most of the flowers brought into the United States come from Central and South America. The climate is favorable to the production of flower crops, and countries in the southern hemisphere can grow flowers during our winter. These are then shipped to the United States. Also, the cost of labor is low compared to other areas.

Other countries also send flowers to the United States. Australia and New Zealand sell their orchids here. Holland has always been famous for brilliantly colored tulips. Each year Holland ships out not only tulip flowers but also tulip **bulbs.** Israel has begun producing roses and is rapidly becoming a leading source for these flowers.

In return, the United States ships gladiolus all over the world. The fresh flower industry is a worldwide business. Many countries have close ties to each other in the floriculture business.

**FIGURE 6-2**
Tulips grown in Holland are shipped all over the world. *Courtesy of Jeff Lewis, The University of Georgia.*

**FIGURE 6-3**
Some flowers need a cool climate and others need a warm climate. *Courtesy of Washington State University.*

# National Aspect

As with the international aspect, flower production in the United States is concentrated in areas where the climate is best suited. Most floral crops need very bright light to grow and produce quickly. Also, some flowers need a cooler climate and others need a warm climate.

The most popular states for growing flowers are Florida and California because of the warm climates. Colorado and the New England states are also heavily involved in producing flowers that require cooler climates. Bulbs, such as tulips and daffodils, need slightly different production climates. Cool summers and mild winters make Washington, Oregon, and California more suitable for producing these.

# Plants for Florists

Many different kinds of plants are produced for **florists.** Everything from blooming plants in pots and hanging baskets to cut flowers such as roses and daisies are available to florists. This is a growing industry in this country. People buy flowers

**FIGURE 6-4**
Creating floral arrangements is a growing business. *Courtesy of USDA.*

for weddings, funerals, anniversaries, birthdays, and other occasions. People also send flowers to cheer up friends who are sick or just to let them know they are special.

## Potted Plants

Any plant grown and sold in a pot is referred to as a **potted plant.** Outside plants such as azaleas and chrysanthemums are grown in pots for sale at the florists. These are usually removed from the pot and transplanted outside. Blooming house plants like kalanchoe and cactus are sold in pots and generally remain in the pots.

Potted plants offer a challenge to the florist. Not only must they have beautiful or unusual foliage or blooms, but the plants have to be healthy. A potted plant, in most cases, must live, bloom, and remain attractive for a long time.

Most potted plants are grown from cuttings while others are started from seed. It can take several weeks to a few months to produce a healthy, blooming potted plant.

**FIGURE 6-5**
Blooming house plants are sold and live in pots. *Courtesy of James Strawser, The University of Georgia.*

**FIGURE 6-6**
Foliage plants are grown for their attractive leaves. *Courtesy of James Strawser, The University of Georgia.*

## Foliage Plants

Plants grown for their leaves are called **foliage plants.** Many times these are grown in pots and could also be called potted plants. Some plants, however, are grown for foliage for flower arrangements.

The most common plant for this use is the leatherleaf fern. It's fronds (leaves) are particularly strong. In many arrangements leatherleaf accounts for most of the plant material. It is used to hide the support structure of the arrangement and as a green backdrop for the colorful flowers.

Another kind of fern, the Boston fern, is popular for hanging baskets. Other plants often used for hanging baskets are begonias and impatiens.

Often several small foliage plants are put together in a shallow pot and sold as dish gardens. Groupings of plants inside glass or plastic containers are called terrariums.

The most popular foliage plants, however, are those grown for homes and public buildings. These can be sold in small pots but are also offered in very large sizes. Landscapers use these to make the yards and outsides of homes, businesses, and other structures more attractive.

**FIGURE 6-7**
Plants can be used in a variety of 's to create beauty. *Courtesy ‑ Strawser, The University*

**FIGURE 6-8**
The Venus flytrap is an exotic plant that catches insects. *Courtesy of James Strawser, The University of Georgia.*

## Exotic Plants

Unusual or rare plants are called exotic plants. Many exotic plants are great for **house plants.** These are attractive because they are so unusual.

An example of an exotic plant is the bromeliads. These plants are unusual because of their cup-shaped leaves. The leaves act as cups because they hold water. For this reason, they can stand longer periods of neglect. Another example is the carnivorous plants. These are plants that eat flesh to grow. The Venus flytrap is the most common carnivorous plant. It traps flies and other insects and digests them. Another carnivorous plant is the pitcher plant. It traps insects by luring them into its "throat." The sides are so slippery the insect cannot get out. It then falls into the digestive fluids in the bottom of the "pitcher" and dies. There is a good market for these exotic plants. People buy them to decorate their homes.

## Cut-Flower Production

The main items in the floriculture trade, of course, are flowers. **Cut flowers** are flowers that are to

**FIGURE 6-9**
Cut flowers are trimmed so they can fit into a vase. *Courtesy of Frank Flanders, The University of Georgia.*

be used in flower arrangements. They are cut at the stems so they can fit into vases and other containers. This differentiates cut flowers from flower-

## Career Development Events

**FFA**

The FFA Floriculture Career Development Event is a competition for teams of four members. Each member competes individually and the final team score is the total of the three highest individual scores. The event consists of four parts: 1) identification of plant materials, 2) placing classes, 3) general knowledge, and 4) practium.

The identification of plant materials consists of plants selected from the State Floriculture list and is displayed for visual identification. Students may not touch the live plant material but can use a magnifying glass to inspect the foliage. The judging classes are plant materials or floral arrangements. The student places the materials from best to worst based on its structural form, market value, and health. Floriculture plants are selected from the identification list and may be cut flowers, foliage plants, bedding plants, or potted plants. The general knowledge part of the floriculture event consists of 50 multiple-choice

questions. This phase of the event tests knowledge and understanding of all areas of the floriculture industry. Students preparing for this part of the event must

The Floriculture Career Development Event tests students' ability to work with floral arrangements. *Courtesy of Carol Duval, National FFA Organization.*

ing plants. Flowering plants may be sold as potted plants. The plants that produce cut flowers are not sold. Only the flowers and stems are marketed.

study not only the operations and maintenance of a greenhouse but also the physiology, growth habits, nutritional needs, disorders, and propagation practices of floriculture crops. Questions addressing floral design, arrangements, equipment, and principles are included as well as customer service, interpersonal relations, salesmanship, and records and reports.

The part of the event known as the practium is where the student identifies plant disorders. Plant disorders include disease, insects, weeds, and nutritional problems. The disorders are displayed for the student to view with or without a magnifying glass. The student is asked to identify the problem and those pesticides or other treatments necessary to remedy the problem. Hands-on practiums are included in the event. Students are asked to perform different skills required in the floriculture industry such as potting plants, designing a floral arrangement, or taking a phone order.

After all team members have completed each phase of the event, each individual score is calculated. The top three individual scores are added to yield the team score. If there is a tie between two team scores, the written exam score is used as a tiebreaker for first and second place teams. State winning teams advance to the National Floriculture Career Development Event.

The FFA Floriculture Proficiency Award is one of the four horticulture proficiency awards. The award is presented to individual FFA members who have excelled in experiences using the principles and practices of field or greenhouse production of flowers, foliage, and related plant materials for ornamental purposes. To apply, the student must be involved in a Supervised Agricultural Experience Program and have maintained a list of floriculture skills he or she has learned. The student must also develop an inventory of buildings, equipment, tools, and materials used in the experience program. He or she must also keep accurate records of income, expenses, and accomplishments. Students may compete within their chapter, district, and state. The state award winners compete for national recognition.

Many nurseries produce flowers just for florists. Many produce only one kind of flower, such as chrysanthemums or carnations. For the growers to make a profit from their crops, they must practice good growing habits and timing.

## Physical Environment

Although some cut flowers, such as gladiolus, are often grown outside in a field, most are grown in greenhouses. There are several reasons for this.

First is temperature. To induce plants to form buds, then flowers, they must have a certain range in temperature. Cold temperatures will keep many plants from forming buds at all. High temperatures may cause them to abort buds and blooms. Temperature can be controlled inside a greenhouse.

Another important factor is moisture. Plants for cut flower production must have just the right amount of water to do their best. Producers can use sprinklers or underground water to keep plants in the field watered. However, it is difficult to prevent the plants from getting too much rain. Hard rain can also destroy the fragile blooms. Just as rain storms can damage plants, so can wind storms. By

**FIGURE 6-10**
In a greenhouse, flower growers can control the environment. *Courtesy of James Strawser, The University of Georgia.*

growing the plants in the controlled environment of a greenhouse, these conditions can be avoided.

## Greenhouse Benches

Benches in greenhouses used for growing potted plants are usually from 5 to 8 feet wide. Benches used for the growing of cut flowers are only 4 feet wide. This is helpful for the intense care these plants need. Plants used for cut flowers are usually not grown in pots. Instead, the benches hold soil and the plants are grown in this. These benches are also closer to the ground. Often flower stalks must be staked or tied to produce straight stems, especially with the large flowers. Lower benches allow more room in which to work.

## Growing Cut Flowers

The grower usually starts with a cutting. Let's follow a typical crop of chrysanthemums. Some growers make their own cuttings, root them, grow them out, induce blooming, harvest the flowers, and ship them. More often, they start with rooted cuttings.

**FIGURE 6-11**
Greenhouse benches hold pots for growing plants. *Courtesy of James Strawser, The University of Georgia.*

**FIGURE 6-12** Some growers start new plants by taking cuttings as shown in (A) and doing rootings as shown in (B). *Courtesy of James Strawser, The University of Georgia.*

**FIGURE 6-13**
Pinching makes the plants produce more limbs. *Courtesy of James Strawser, The University of Georgia.*

This is where timing comes in. Growers usually have several crops growing at once so they can have continuous production. A new crop is planted every two to four weeks. Different types of chrysanthemums mature at different times, so the grower must take this into account also.

The cuttings are planted in the low, narrow benches and grown out. Wire mesh, usually with 6-by 8-inch spacing, is laid on top of the soil before planting. This is raised as the plants grow to assure straight stems on the flowers.

About two weeks after planting the plants are pinched. This refers to the practice of taking off the bud at the very tip of the plant. The plant will then produce several stems from buds farther down the stem. All but two or three of these stems are taken off. This makes the plants produce more flowers.

## Lighting

Chrysanthemums are **photoperiodic.** This means that they respond to the length of night. Only during long nights, or short days, do chrysanthemums produce flower buds. Left alone, they flower only during the fall and early winter. To get around this, the dark period is artificially controlled. To allow the plants to grow for a while without forming buds, the dark period is interrupted with electric lights.

When the plants are larger and the stems are long enough, lighting is no longer as important. If the greenhouse is in a rural area, without outside light coming in and it is the right time of year, the plants will form flower buds naturally. Most of time, however, this is not left to chance. The plants are covered each evening with a solid piece of black cloth that blocks all light. The cloth is then removed in the morning. It only takes a very small amount of light from almost any source to inter-

**FIGURE 6-14**
A cloth over the greenhouse can help control the amount of light the plants get. *Courtesy of James Strawser, The University of Georgia.*

rupted the cycle. This continues until the flowers are well developed.

## Temperature

Temperature must be monitored frequently. Although different cut flower plants need slightly different temperatures, chrysanthemums need a night temperature of 60 to 62 degrees and a day temperature of 65 to 70 degrees.

If temperatures go above 80 degrees or so, it takes longer for flowers to be produced and the flower quality is lowered. On the other hand, if temperatures are too cool, chances of disease go up, production time is longer, and the flower size and stem length are both less.

## Timing

The grower must follow a precise timetable. This timetable must be started from the date the flowers must be ready and worked backward to pinpoint when the plants should be started, when short or long days should be started, when the plants

**FIGURE 6-15**
Poinsettias require rigid schedules of lighting, watering, and temperature control. *Courtesy of James Strawser, The University of Georgia.*

should be pinched, when fertilized, and when temperatures should be controlled.

Timing is more important with some flowers. For example, unlike chrysanthemums, which are used all year in flower arrangements, poinsettias are grown for and bought only at Christmas. Green poinsettias are unsalable in December. Poinsettias that are a beautiful red and green on December 26 are likewise useless. Rigid schedules of lighting, watering, and temperature are used to make them bloom at the right time.

## Harvesting

Chrysanthemum flowers may be harvested while they are just opening or in full bloom. This depends on the type of flower and how far the flowers will travel to market. The stem is cut as long as it can be. This is to give the florist enough stem length for a variety of uses. Harvesting continues for one to two weeks. Then the plants are discarded and the bed is prepared for the next crop.

The flowers are immediately put into room-temperature water, then into a cooler where they are kept at a much lower temperature. Once the flowers are cut, they are no longer living and will begin to decay. Cool temperatures delay the decay until the flowers are used. Later the flowers are graded. The best grade includes the largest, most perfect flowers. Smaller flowers, or slightly off-color flowers are lower grades.

## Transportation

Flowers are easily damaged and special measures are taken to protect them. The bunches (for chrysanthemums there are usually 12 in a bunch) are wrapped in a clear plastic. Large, sturdy cardboard boxes are used to further protect the flowers. Newspaper is packed gently around the flowers to keep them from jostling.

Flowers are transported in refrigerated trucks for short distances. For longer distances, such as from California to Maine, or Colombia to Miami, they are shipped in refrigerated compartments in airplanes.

Some of the more delicate flowers only last a mater of days, then are unusable. For this reason, the florist must know when to order these.

**FIGURE 6-16**
The best grade of flowers is large and perfect. *Courtesy of James Strawser, The University of Georgia.*

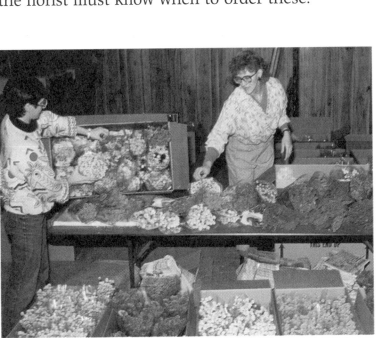

**FIGURE 6-17**
Flowers have to be carefully packed to prevent damage. *Courtesy of University of Minnesota Agricultural Experiment Station.*

**FIGURE 6-18**
A nice floral arrangement must have a focal point, balance, scale, and unity. *Courtesy of James Strawser, The University of Georgia.*

# Flower Arrangements

How well the flowers look depends to a large degree on how they are presented or arranged. Among the principles of flower arranging are balance, scale, focal point, and unity.

Balance refers to the apparent weight of the flowers. An arrangement should not look like it is about to fall over. By balancing large flowers and light and dark flowers the designer creates a beautiful piece that looks stable.

Scale involves the various sizes of flowers compared to each other and the container. It would not look right if the arrangement consisted of a few small flowers in a very large vase.

An arrangement should have something that draws the viewer's attention. This is called the focal point. The focal point is usually in the front, near the bottom of the arrangement. It can be a very large flower, a very dark flower, or an unusual flower.

The design should flow together. A wild flower bouquet with one rose would be out of place. Unity is necessary in an arrangement. All the parts should fit as one.

# Summary

From growing flowers and plants to selling the finished product, floriculture is a worldwide business. It takes planning and forethought to produce and sell floriculture products.

# 6 CHAPTER REVIEW

## Student Learning Activities

1. Bring a variety of different types of flowers to class. Arrange them in different ways. Try to identify several principles that apply to making floral arrangements that look good.

2. Interview a florist. Determine where the flowers used by the florist come from and the main markets for the arranged flowers. Share your findings with the class.

3. Obtain a potted plant from a department store, florist, or a friend. Go to the library and conduct research about the plant. Find out the country where the plant originated and where the domesticated plants are grown. Also determine how the plants are produced. Report to the class.

## True/False

1. ___ Most potted plants are started from seed.

2. ___ Floriculture is the growing of flowers.

3. ___ Plants grown for their leaves are called foliage plants.

4. ___ Photoperiodic means that the plants respond to the length of the day.

5. ___ If temperatures rise above approximately 80 degrees, it takes longer for flowers to be produced and the flower quality is lowered.

6. ___ Balance refers to the apparent weight of the flowers.

7. ___ The most common plant grown for foliage is the pitcher plant.

8. ___ Dish gardens are deep pots for small foliage plants.

9. ___ Unusual or rare plants are called exotic plants.

10. ___ Terrariums are plants growing inside a glass or plastic container.

# Multiple Choice

1. Most of the flowers brought into the United States come from
    a. China and Japan.
    b. Central and South America.
    c. Australia and New Zealand.

2. One of the leading sources for roses is
    a. China.
    b. Israel.
    c. Australia.

3. Flower production in the United States is greatest in
    a. Florida and California.
    b. Georgia and Florida.
    c. North Carolina and California.

4. The main items in the floriculture trade are
    a. shrubs.
    b. plants.
    c. flowers.

5. Benches in greenhouses used for growing potted plants are usually between
    a. 5 and 8 feet wide.
    b. 3 and 4 feet wide.
    c. 1 and 2 feet wide.

6. The most obvious example of the importance of timing involves
    a. poinsettias.
    b. chrysanthemums.
    c. leatherleaf fern.

7. About two weeks after planting, the plants are
    a. graded.
    b. pinched.
    c. cut.

8. Cut flowers are used for
    a. flower arrangements.
    b. foliage.
    c. hanging baskets.

9. The Holland is famous for its
    a. daisies.
    b. roses.
    c. tulips.

10. Fronds are another name for
    a. buds.
    b. stems.
    c. leaves.

# Discussion

1. Why are most cut flowers grown in greenhouses?

2. What does scale mean?

3. Where do most flowers bought in the United States come from and why?

4. Why are potted plants a challenge to florists?

5. How does a pitcher plant trap insects?

6. How can temperature, rain, and wind affect plants?

7. Why is floriculture such a big business throughout the world?

8. How are cut flowers grown?

9. Why are poinsettias such a good example of timing?

10. Give some examples of exotic plants.

# Nursery Production

*When you have finished studying this chapter, you should be able to:*

- Define a plant nursery.
- Describe materials nurseries use.
- Identify the basic greenhouse types.
- Describe what a mist system does.
- Distinguish sexual from asexual propagation.
- Describe the importance of breeding programs.

H ave you ever wondered where all the flowering plants, shrubs, and trees around our houses come from? They come from **plant nurseries.** People who work at nurseries start with seeds or pieces of plants and grow them out to whole plants. These plants are sold in everything from 1-inch pots to 50-gallon pots. Some are even grown directly in the ground, then dug up when sold. Some of these plants must be moved with huge trucks due to their size.

Some growers produce their own plants, grow them out then sell them. In

FIGURE 7-1
The plants around your home prob-
ably originated at a nursery.
*Courtesy of James Strawser, The
University of Georgia.*

some nurseries there are entire teams of people who only take cuttings. These people are very skilled and can take up to 8,000 cuttings in a single day.

After these cuttings are taken from the **stock plants** (plants grown for the purpose of producing cuttings), they are treated with a chemical that induces rooting, then placed in soil in greenhouses.

Many nursery businesses buy cuttings or small plants (called liners) and grow them to sell. The liners are potted up in larger containers and

FIGURE 7-2
Cuttings from superior plants are
used to start new plants. *Courtesy of
James Strawser, The University of
Georgia.*

allowed to grow for a few months to a year. After the plants have filled out the larger container they are either sold or transplanted into larger containers that will be sold later.

Plants bought from other sources for the grower to work with should arrive quickly and be healthy. Unrooted cuttings do not survive long. They must be handled very carefully. Frequently they are put into a cooled room until they can be processed.

## Growing Media

The substances that plants are grown in are called **media.** Why not call it soil? Becuase many plants are not grown in soil. Over the last several years growers have begun to rely more and more on "soilless" media.

Media must combine several characteristics. They must be well drained so the roots of a plant don't rot. But at the same time, they need the ability to retain water for the plant's use.

Soilless media are usually made of a mixture of ingredients. A basic component can be peat. Peat is compressed organic matter, usually harvested from bogs that have been drained. Another common ingredient is pine bark that has been crushed into small pieces. Sand is added in some instances.

Human-made elements include vermiculite (which is actually heat-expanded rock), rock stuff, and perlite. These artificial ingredients have the advantage of being lightweight. This is especially important for nurseries that ship their plants using freight companies that charge by weight.

Containers used in the nursery industry vary widely. Many seeds are started in flats. These are trays filled with media. Usually fine, lightweight media are used to make it easier for the seedlings to grow.

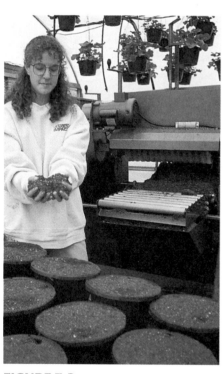

**FIGURE 7-3**
Soilless media is mixed from a number of ingredients. *Courtesy of James Strawser, The University of Georgia.*

**FIGURE 7-4**
Seeds are usually started in flats. *Courtesy of James Strawser, The University of Georgia.*

**FIGURE 7-5**
Bedding plants are planted and sold in nursery cell packs. *Courtesy of James Strawser, The University of Georgia.*

**Bedding plants** (plants, usually annuals, that are planted in large quantities) are usually sold in cell packs. Cell packs come in several sizes, from three packs to the new larger jumbo packs. These are convenient for the person purchasing the plants. The plants are taken from the cells and planted.

Many plants are started in 2.5 inch pots. After growing in these pots, they are usually transplanted into larger pots. Some are produced as liners to be sold to other nurseries.

For very large trees and shrubs, containers as large as 30 to 50 gallons are sometimes used. Large plants grown in the field are dug up and burlap is wrapped around the root system. This helps keep the roots moist and bound together to prevent injury. The burlap can then be planted with the tree. It decays in a short time.

Most growers use plastic pots, although a few use clay. Clay pots are much heavier than plastic ones and are easier to break. Clay pots are also porous,

**FIGURE 7-6**
Large plants are dug up and the roots are wrapped in burlap. This is called balled and burlapped. *Courtesy of James Strawser, The University of Georgia.*

meaning they have tiny holes throughout them. This ensures that the plant roots will get oxygen and avoids overwatering problems. On the other hand, because the clay pots are porous, water is lost faster and the plants must be watered more often.

The color of the pots is sometimes important. Black pots tend to absorb more of the sun's energy and keep the roots warm. This is good in winter, but it is sometimes a problem in summer. White pots keep the roots cooler, but they break down in the sunlight and do not last as long as black plastic pots. A few nurseries color-code their crops. Blue, pink, orange, and other colors may signify the time the plant was repotted, the type of plant, or the color of the blooms.

## Growing Areas

Most nurseries have both inside and outside growing areas. Both the climate and the plants grown have an effect on the amount of outside and inside growing areas. Plants like daylilies can be grown in fields. House plants, such as poinsettias for Christmas must be grown in covered, heated areas.

FIGURE 7-8
Most greenhouses are covered in fiberglass or plastic. *Courtesy of James Strawser, The University of Georgia.*

Greenhouses are the most common enclosed growing areas. These structures allow light in and protect plants from the weather. Greenhouses are usually heated. This allows the grower to extend the growing season and produce plants all year long. Moisture can be precisely controlled on greenhouses. Plants can be given water only when they need it. This protects the plants from overwatering during very wet weather.

Most modern greenhouses are made of fiberglass or polyethylene. Fiberglass is ridged material attached to a metal or wood frame. Fairly inexpensive, fiberglass weighs less than glass, but yellows with age, cutting down on the amount of sunlight that gets through. Polyethylene resembles large sheets of plastic and is even less expensive than fiberglass. However, it only lasts one to five years and then must be replaced. These large pieces of polyethylene are also attached to a metal or wooden frame.

The shape and size of the greenhouse used depends on what the nursery is producing and where it is located. In the North, the greenhouse must be able to withstand snow loads. In the South, it is more important that the greenhouse be easily cooled.

Most of the modern nurseries are putting in Quonset greenhouses. These look like a barrel cut in half. Almost all of these are built to be covered in polyethylene, although a few are manufactured for fiberglass coverings.

Freestanding even- or uneven-span greenhouses are better able to take snow loads. Some of these structures are covered with glass, but most are covered with fiberglass.

In larger nurseries several greenhouses are connected at the sides. This is called ridge and furrow. It allows for several different climates inside, while still being more cost efficient as far as heating and cooling goes.

A group of greenhouses at a nursery is called a range. A range can be made of all one type of greenhouse, such as Quonsets, or a mixture of types, such as Quonsets and freestanding or even-span greenhouses.

Some greenhouses are put up just to shield plants from wind and very low temperatures. These greenhouses have no supplemental heat in them. Because of this, they are called **cold frames.** Cold frames can be small structures that are mobile and just laid over plants in a bed, or they can be very large with benches just like heated greenhouses.

**Cold Frame**

FIGURE 7-10
Cold frames have no artificial heat and only provide shelter for the plants.

# Outside Growing Areas

The most common outside growing structure is usually a lath house. A lath house is a metal, or more commonly, a wooden structure with **shade cloth** stretched over the top and sometimes sides. Shade cloth is loosely woven black cloth that blocks part of the sun. These black cloths are commercially available in weaves that block from 5 to 90 percent of the sunlight.

Unlike greenhouses, water is not controlled and temperature is only changed a little. These houses are used only to provide shade to those plants that do not like a lot of sun.

Some nurseries use outside beds to grow out plants. Treated boards are used to form raised beds. These are then filled with sand, soilless media, or plain soil. Plants are grown in them, then sold bare root, or without a pot. Sometimes these plants are dug up and then potted into containers for sale.

Some areas outside are covered with landscaping cloth. Landscaping cloth is a thick, black cloth

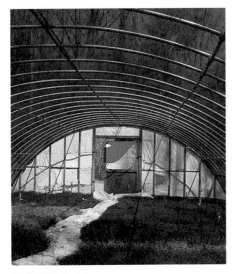

FIGURE 7-11
Shade cloths partially block out the sun. *Courtesy of James Strawser, The University of Georgia.*

# Career Development Events

**FFA**

The Nursery Operations Proficiency award is an individual competition. FFA members who have job experience with turf, bedding plants, shrubs, and/or trees for production operations may apply for the Nursery Operation Proficiency. To apply, the student's supervised Agricultural Experience Program (SAEP), school classes, and FFA activities should be related to this area. Isabel was the winner of the chapter proficiency award in nursery operations. She had spent over 1,800 hours working for a local grower. Isabel's work experience included propagation of shrubs, pruning field-grown trees, fertilizing container stock, and application of chemicals and fertilizer. Isabel also developed skills in communication and computer applications. She was active in her FFA fundraising and community awareness activities and served as chapter secretary for two years. Isabel's accurate recordkeeping and wide range of experience and activities made her an excellent candidate for a proficiency award. All proficiency awards are judged on the students' application and experience. Good recordkeeping plus the development of proficiency-related skills and accom-

plishments are essential to become a winner. Students may compete within their chapter, district, and state. The state winners compete for national recognition.

Another opportunity for individuals to be awarded through their interests in the nursery industry is the FFA Achievement Awards. Achievement Awards are lists of skills, abilities, and attitudes needed to enter a career in the area of the student's interest. The Ornamental Horticulture Achievement Award is divided into four areas: Career Exploration, Career Skills, Leadership Development, and Safety Practices. Each of these four areas has a list of skills and abilities, and the student has to master 80 percent of each list to receive the award. The agriculture instructor or even a worksite employer can evaluate and check the student's accomplishments.

Eddie wanted to learn more about the nursery industry and set out to win the Ornamental Horticulture Achievement Award. Eddie first worked on accomplishing the Career Exploration skills by visiting and interviewing five different horticulture operations and presenting a written report to his teacher. He also had to learn the skills required to compete as a team member in the nursery/ landscape career development event and participated in the FFA chapter tree planting program. To satisfy the Career Skills, Eddie prepared a scrapbook. The scrapbook identified 100 plants, nursery media, pruning techniques, insect and disease problems of 20 landscape plants and a list of tools and equipment

FFA members who enjoy working with plants can get involved with projects in nurseries and apply for the Nursery Operations Proficiency Award. *Courtesy of Frank Flanders, The Agricultural Education Program, The University of Georgia.*

needed in a nursery operation. Next Eddie had to demonstrate to his teacher the proper ways to propagate a shrub, plant and prune a tree, apply fertilizer, spray insecticide, and keep records of his plants in the school greenhouse. Eddie obtained his points in Leadership Development by attending chapter meetings, participating in chapter fund-raising activities, serving as chapter chaplain, and entering the public speaking contest. Eddie's employer at the landscape business checked his knowledge of safety practices. Eddie demonstrated the safe use of tools and equipment used in his job. After Eddie completed 80 percent or more of the skills as verified by the evaluator, he was presented with the official FFA Achievement Award Certificate at the school's award program.

All FFA members should strive to earn at least one Achievement Award. After deciding upon your career interest, work with your parents and advisor to select the skills and experiences you want to attain.

that forms a barrier to weeds and holds down erosion. Containers are put directly on the landscaping cloth and grown out, then sold.

Another area that is used for growing plants outside is the ground itself. Plants are put into the ground, allowed to reach a good size, then harvested and sold bare root.

## Mist Areas

Some plant reproduction techniques involve the use of mist. Tiny droplets of water from irrigation lines keep the plants moist so they don't use up their water reserves. There are two main types of mist systems used. The first type is a timed system. A solenoid wired into the irrigation lines controls the flow of water. The solenoid is controlled by a timer. The water goes on for a short time (a few seconds) every five minutes to every hour, depending on circumstances.

The second type uses a mechanical leaf. A piece of wire is attached to an electrical device. When the screen (the mechanical leaf) is wet, the weight of

**FIGURE 7-12**
Irrigation systems in modern greenhouses are timed to water at certain times. *Courtesy of James Strawser, The University of Georgia.*

the water on the screen tilts down, trips the sole-noid, and shuts off the water. As the screen dries out it becomes lighter. When this happens, it tilts up and trips the solenoid to let water through. The water makes the mechanical leaf heavy again and turns the solenoid off.

# Propagation

Plant reproduction, or making several plants from one plant, is called **propagation.** This is done sev-eral ways, depending on the plant and the nursery operation.

Some operations produce their own seed and grow plants from them. Others buy seed to begin plant production. Some only handle pieces of plants. Others may combine the two main propagation methods.

## Sexual Propagation

Sexual propagation refers to producing plants from seeds. Plants have been producing seeds for many millions of years. Humans began taking advant-age of this at the dawn of civilization. Plants that they found useful were propagated to produce new plants.

Sexual propagation begins with the flower. Flowers differ in size, color, and parts. Some plants have only female flowers on them, others only male. Other types of plants have male and female flowers separately on the same plant. Most plants, however, produce flowers that are both male and female.

The female parts of a flower, called the pistil, consists of the stigma, the style and the ovary. Egg cells are produced in the ovary.

The male parts of the flower, called the stamen, contain the anther and filament. The anther pro-duces pollen, or sperm cells. The pollen, carried

**FIGURE 7-13**
Annual plants are often started from seed. *Courtesy of James Strawser, The University of Georgia.*

by the wind or insects, lands on the stigma. This is called pollination. It then forms a pollen tube through the style to the ovary. When the sperm cell actually unites with the egg cell fertilization has occurred.

Flowers fertilize each other or themselves to produce seeds. When they fertilize themselves they are self-pollinating. When they pollinate another flower they are cross-pollinating. When the flowers pollinate the male and female cells are united. This forms the zygote that develops into the seed.

Many nurseries that produce annual bedding plants start with seed they order from another nursery. They germinate the seed, then grow it out. Germination is when a seed starts to grow into a plant.

Growing plants from seed is less expensive than growing plants from cuttings. It enables a grower to produce many thousands of plants quickly.

## Asexual Propagation

Sometimes it is less desirable to start plants from seed. Especially for large trees and shrubs, it takes longer to get a salable plant starting from seed. In some instances, a plant doesn't "come true" from seed. That means that the plant from a seed may not be exactly like the parent plant. This is because of cross-pollination where characteristics are obtained from both parent plants.

For this reason, many growers produce their plants using asexual propagation. This method starts with a part of a plant itself instead of a seed produced by the plant.

## Cuttings

The most popular form of asexual propagation is taking **cuttings** from the plant. A piece of the stem (or sometimes a leaf) is taken from the plant. It is usually treated with a rooting hormone (a chemi-

**FIGURE 7-14**
Root hormones are applied to help a cutting grow new roots. *Courtesy of James Strawser, The University of Georgia.*

cal that makes the plant form roots), then placed in media. The cutting is then placed under mist until it has formed roots and no longer needs the extra water.

Propagating plants this way assures the grower of a crop that will be almost exactly like the original. In addition, the plants will be uniform in growth. This is especially important for landscapers who buy many plants for hedges or beds.

Some plants cannot be propagated this simply. In this case, other asexual methods are used. Some plants are propagated by air-layering. The stem is peeled all the way around, a rooting hormone is painted on the wound, and the area is then wrapped in damp media, covered with plastic and allowed to form roots while still getting water from the parent plant.

Other plants must be propagated by root cuttings. The roots are cut off much as the stems are, then allowed to produce a plant.

## Tissue Culture

Micropropagation, or tissue culture, produces an entire plant from a single cell of the parent plant.

**FIGURE 7-15**
Many plants can be propagated by separating or dividing the roots or bulbs. *Courtesy of James Strawser, The University of Georgia.*

**FIGURE 7-16**
Holly is valued for its bright leaves and berries. *Courtesy of James Strawser, The University of Georgia.*

Although it takes quite a bit longer than producing plants from cuttings, it has the potential to produce millions of plants from a single branch of a plant.

Still other plants must rely on **division** or separation for propagation. This involves digging up part of a plant just below the soil surface and separating it into several parts. Some plants can be gently pulled apart, called separating. Other plants must be cut apart into sections; this is called dividing. These parts are then allowed to grow into whole plants.

## Plant Production

Let's follow a typical plant from establishment to selling. One of the most popular plants sold in nurseries today is the holly. Hollies are valued for their glossy green foliage, red berries, and screening ability. A row of hollies can screen a busy highway or noisy neighbors. They are common in landscaping plans.

Propagating *Ilex crenata,* the Japanese holly, begins with stock plants. These are large plants that are either planted on the nursery grounds to be used as cutting material, or older plants in large containers.

The cuttings must be taken from the right part of the plant. Branches that look sick or were heavily shaded by the rest of the plant should be avoided. Old branches with thick bark should also be left. Very young, weak tips may rot under mist. Usually young, firm wood is taken. The cuttings are between 4 and 6 inches long.

The best time to take cuttings is in the cool morning. If exposed to heat and direct sunlight, the cuttings may suffer enough damage to be unusable. Cuttings are kept in a cool, shaded place and kept moist until ready to process. Depending on the operation, a few hundred cuttings may be taken, or several thousand before they are all processed.

The lc ver leaves are taken off to expose the bottom node. Often the bark is scraped off in a small area to help the plant to form roots. A powdered or liquid rooting hormone is then applied to the bottom of the stem.

The cuttings are then stuck into moist media in small pots, bare root trays, or outside beds under mist. Keeping the new cuttings moist is especially important. If allowed to dry out too much, the cuttings will die. If the area is kept too wet, fungus and bacteria may become a problem.

The cuttings stay there until they form roots. They are misted less and less frequently. After three to four weeks the cuttings are weaned off the mist.

Once the root system has filled the small pots, they are repotted into 1-gallon containers or sold as liners. When the roots have filled the 1-gallon pots they are either sold or repotted again. Hollies are available in as large as 50-gallon pots. This size is rather expensive, however, and the majority of hollies are sold in 1- to 3-gallon pots.

While growing out, the hollies are fertilized, either with constant liquid fertilizer, or a granular fertilizer put on one to three times a year. They are sprayed for pests, and the pots kept weed free. Water is used when needed, usually twice a week or so, depending on weather conditions.

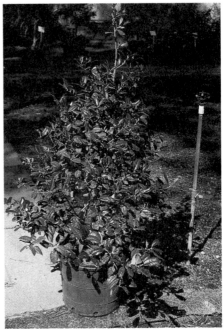

**FIGURE 7-17**
Young holly plants are sold in pots.
*Courtesy of James Strawser, The University of Georgia.*

## Summary

There are many things to consider when thinking about the nursery business. Nurseries vary widely in the plants they grow and the methods they use. The nursery business is complex and fascinating.

# 7 CHAPTER REVIEW

## Student Learning Activities

1. Take an inventory of the plants around your home or school. Determine which of these originated in a nursery.

2. Visit a nursery in your area. Find out which plants sell the most. Think of some reasons why these particular plants are the most popular. Report to the class.

3. With your teacher's assistance, take cuttings from a plant and root them. Grow the plant in a pot and at the proper time plant it at your home.

## True/False

1. ___ Plants are the basis of the nursery industry.

2. ___ Unrooted cuttings survive a long time.

3. ___ Vermiculite is a human-made element.

4. ___ The most common outside growing structure is a greenhouse.

5. ___ Landscaping cloth is used solely to keep insects away from plants.

6. ___ The most popular form of asexual propagation is taking cuttings from plants.

7. ___ Micropropagation produces an entire plant from a single cell of the parent plant.

8. ___ The best time to take cuttings is in the afternoon.

9. ___ Stock plants are grown for the sole purpose of producing cuttings.

10. ___ All substances that plants are grown in are called media.

# Multiple Choice

1. Human-made elements include
     a. peat.
     b. bark.
     c. perlite.

2. The most common enclosed growing area is the
     a. nursery.
     b. greenhouse.
     c. lath house.

3. Most modern greenhouses are made of
     a. plastic.
     b. metal or wood.
     c. fiber glass or polyethylene.

4. Greenhouses put up to shield plants from wind and temperatures are called
     a. cold frames.
     b. ranges.
     c. lath houses.

5. Shade cloth is used to
     a. kill weeds.
     b. block out the sun.
     c. hold down erosion.

6. Sexual propagation refers to producing plants from
     a. seeds.
     b. leaves.
     c. cuttings.

7. Egg cells are produced in the
     a. ovary.
     b. pollen tube.
     c. stigma.

8. When flowers pollinate another flower it is called
    a. self-pollination.
    b. cross-pollination.
    c. fertilization.

9. Freestanding even- or uneven-span greenhouses are better able to take
    a. water loads.
    b. soil loads.
    c. snow loads.

10. The most common outside growing structure is the
    a. greenhouse.
    b. lath house.
    c. range.

## Discussion

1. Why are plastic pots more common than clay pots?

2. Why is the color of plant pots important?

3. What are the two main types of mist systems?

4. What is the difference between sexual and asexual propagation?

5. List the male and female parts of a flower.

6. What is air layering?

7. What are media?

8. What is a lath house?

9. What is the difference between self-pollination and cross-pollination?

10. When is the best time to take cuttings? Why?

# 8

# Landscaping

## Student Objectives

**When you have finished studying this chapter, you should be able to:**

- Describe the job of a landscape architect.
- List the land and plant requirements a designer must know before installing plants.
- Describe the difference between deciduous and evergreen landscaping plants and list their uses.
- List and describe the five principles of landscape design.
- List the steps in installing plant material.
- List the responsibilities of a landscape maintenance person.
- Describe what an interiorscaper does.

A multimillion dollar business in the United States is **landscaping.** It employs thousands of people in many different areas. So what exactly is landscaping? Simply put, landscaping makes an area (home, business, or park) more beautiful. It involves the use of natural materials, like plants and rocks as well as human-made features, such as walkways, walls, and pools. The idea is to take all of the

**FIGURE 8-1**
Landscaping is a multimillion dollar
business in the United States.
*Courtesy of James Strawser,
The University of Georgia.*

objects on the outside of a building or of an area and make them fit together in an attractive appearance.

A lot of landscaping plants and materials are used by homeowners. Most, however, are used by landscaping companies that are involved in very large projects, such as housing developments, parks, and public buildings. There are three main areas in landscaping:

design
installation
maintenance

## Design

Design is one of the most important parts of landscaping. If the design for an area is not right, the beauty of the flowers, **shrubs,** and other things cannot be fully appreciated.A person who makes landscape designs professionally is called a **landscape architect.** Landscape architects must be able to work well with other people and be knowledgeable in subjects such as communication (both written and oral), math, and basic horticulture.

There are several styles of landscaping design. There are many sizes, shapes, and colors of plants that can be used in landscaping. Some people prefer rough stone walls, while others want neat brick walls. Some want walkways and neatly laid out designs while others may want a more natural look. Whatever the style of design, there are principles of landscape that make the area more attractive.

## Design Principles

There are five basic design principles to take into account: **scale, balance, simplicity, focal point,** and **unity.** Each one is an important part of the overall design for a workable landscape. This is where the artistic ability of the landscaper comes in. The final plan must be neat, attractive, and tailored to fit the individual.

Plant material and human made structures should be in scale with each other. A large bed of low-growing juniper should not have a huge oak tree growing in the middle of it. A very large house should have large trees and bushes around it, not dwarf Japanese maple trees. A small house, however, should not be covered by a huge Leyland cypress. The plants should be in proportion to each other and appear as though they belong with each other.

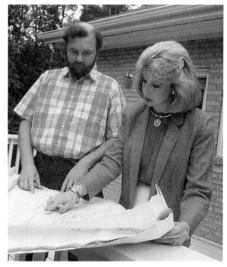

**FIGURE 8-2**
A landscape architect designs landscapes to enhance homes and other buildings. *Courtesy of James Strawser, The University of Georgia.*

**FIGURE 8-3**
Plant materials and human-made structures should be in scale with one another. *Courtesy of Terri Hamlin, Georgia Department of Education.*

When installing a landscape, a designer will often start with plants in 1- to 3-gallon containers and perhaps a few large trees. Keep in mind that these plants will grow. Some may grow very large, even though they may be small when planted. The landscape architect visualizes how the plants will look at maturity.

## Balance

A good landscape design should be balanced. For instance, if on one side of a house there is a large tree, there should be something else on the other side that seems to have the same amount of weight. It does not necessarily have to be another tree. It could be a grouping of shrubs that added together seems to have the same mass as the tree.

A large shrub at one corner of the house with a short ground cover plant on the other corner would be unbalanced. In the same way, a thin, loose-limbed shrub would be unbalanced by a thick, dark-leafed shrub.

Not all landscapes are balanced and not all clients particularly want their yards balanced. Balance is a type of formal placing of plants. For

**FIGURE 8-4**
A good landscape design should be balanced. *Courtesy of W. T. Smith.*

FIGURE 8-5
Designs should be kept simple.
*Courtesy of James Strawser, The University of Georgia.*

clients that want a more casual look, the landscape does not have to be balanced.

## Simplicity

Having one of many different kinds of plants in a landscape will make it look cluttered. Instead the design should be kept simple. Several of the same type of plant should be used together and perhaps other types in another area in a mass planting.

Instead of having every hue in the rainbow in flower borders, limit them to two or three basic colors. Many famous garden designers restrict themselves even more. Gertrude Jekyll is a famous landscape artist. She has stressed this landscaping principle in her famous borders. One border flower garden contains nothing but blue and white flowers and is acknowledged as one of her most beautiful works.

Simplicity is also important in the shape of flower and plant beds. The beds should have smooth, rounded shapes, not a grouping of bulges and indentions. Simplicity lowers the time needed to maintain the landscape. It also makes the area appear more tranquil and peaceful.

FIGURE 8-6
A feature should stand out as the focal point. *Courtesy of Neal Weatherly, The University of Georgia.*

## Focal Point

There should be something in every area of the landscape that stands out or draws attention. This is called the focal point. A focal point can be an unusual plant, a statue, pool, or spectacular view. This feature should not be overdone. If half a dozen focal points are mixed into each area, the eye will not move slowly over the landscape and the impact of these focal points will be lost to the viewer. The principle of simplicity is also lost.

In some areas the focal point may be the house itself. A porch with rockers, a large bay window, or the front door may be a focal point. The idea is to centralize the landscape around the focal point.

## Unity

The last principle is sometimes the hardest to master. A landscape should flow together or look like it belongs as a whole. This is unity. A quiet simple Japanese garden style, for example, should not be interrupted with the stand out, bright colors of a fuchsia plant. Rather the same style should prevail.

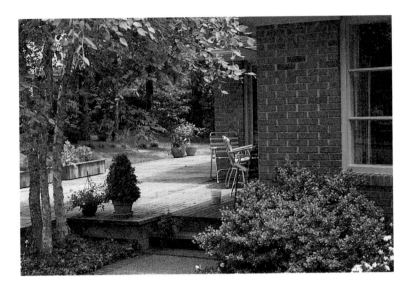

FIGURE 8-7
A landscape should flow together and look like it belongs together. This is called unity. *Courtesy of Neal Weatherly, The University of Georgia.*

Unity is also expressed by using the same plant several places throughout the design. For example, rosebushes used sparsely through a garden give a sense of continuity.

A brick or stone pathway can create unity as it ambles from one area of the landscape to another. The use of the same edging material, such as landscaping ties or rocks can do the same thing. These materials give the illusion of tying the landscape together.

## The Plan

Before the landscaping is begun, all of the design is sketched out in a plan. The plan should be clear and easy to read. It should include a list not only of the plants wanted, but also the size needed (for example a 5 foot luster leaf holly). Walkways, walls and structures should be clearly marked and described.

A clear idea of how the landscape will look should appear on the landscape design plan. It should be detailed and practical so it can be used to design the work.

## Career Development Events

**FFA**

The FFA Nursery/ Landscape Career Development Event is a competition for teams consisting of four members. Each member competes individually and the final team score is the sum of the three highest individual scores. The contest consists of four parts 1) identification of plant materials 2) placing classes, problem solving, and equipment 3) general knowledge, and 4) practium.

Identification of plant materials consists of plants selected from the State Nursery/Landscape list and displayed for visual identification. Students may not touch the live plant material but can use a magnifying glass. The placing classes, problem solving, and equipment phase of the contest consist of five items with any combination of judging classes, problem solving, and questions about landscape/nursery equipment. The judging classes are plant materials or landscape plans in which the student places the materials from best to worst. Plant material is placed based on its structural form, market value, and health. The landscape plans should be placed by the criteria of design principles, plant selection and placement, and use of basic drafting methods. Nursery/landscape plants will be selected from the identification list and will be the same species and cultivar in each class. Plants are placed by structure, market value, and health.

The general knowledge part of the nursery/landscape event consists of 50 multiple choice questions. This phase of the event tests for knowledge and understanding of all areas of the nursery/ landscape industry. Areas for testing include plant materials, plant disorders, cultural practices, landscape design and construction, supplies and equipment, safety, equipment operation and maintenance, interpersonal relations, salesmanship, and records and reports. In the practium, students identify plant disorders and answer questions about a given landscape plan. Plant disorders include diseases, insects, weeds, and nutritional disorders. Disorders are displayed for the student to view with the naked eye or magnifying glass. The student is asked to identify the problem and those pesticides or other treatments necessary to remedy the problem. The landscape plan practium tests the student's ability to read a landscape plan and knowledge of landscape design principles by answering multiple choice questions.

After all team members have completed each phase of the event, each individual score is calculated. The top three individual scores are added to yield the total team score. If there is a tie between two team scores, the written exam score is used as a tiebreaker for first and second place teams. State winning teams advance to the National Nursery/Landscape Event.

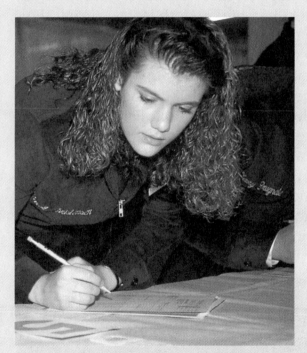

As part of the Nursery/Landscape Career Development Event, students must determine where plants are to be used in landscaping. *Courtesy of Carol Duval, National FFA Organization.*

The Turf and Landscape Management proficiency award is an individual competition. This proficiency award can be applied for if an FFA member's job experience has focused on the installation, maintenance, and design of landscape plants and materials. Vincent is a senior and the winner of the chapter proficiency for turf and landscape. He started his own lawn care service when he was in the eighth grade. He now has contracts with several local businesses and many homeowners. Skills such as mowing, aerating, dethatching, fertilizing, and weed control are part of Vincent's business. Last year he made over $4,000 before his expenses were subtracted. Vincent was on the FFA senior high parliamentary procedure team, livestock show team, and served as chapter vice president. All proficiency awards are judged on the student's application and experience. Good record keeping of inventories and expenses plus the development of proficiency-related skills and accomplishments are essential to be a state and national winner.

# Landscape Maintenance

Once landscapes are installed, they must be maintained to keep them attractive. A multimillion dollar business has grown around maintaining landscapes. Many times a landscaping company or an individual offers landscape maintenance. This means that at specified intervals the company will mow the lawn, weed the beds, make sure the irrigation is working properly and keep an eye out for pest problems.

Probably the most time and money spent in landscape maintenance is on lawn care. Think of some of the large houses you have seen. Many of them probably had big, smooth lawns surrounding them. Keep this in mind: they don't grow and stay that way by themselves.

Lawn maintenance begins in the early spring and continues until late fall. It begins with a general cleanup followed by evaluation. For small areas a maintenance person uses a simple hand rake to remove old grass clippings and dead leaves from trees and shrubs. For large lawns, like those of businesses or large residences, large equipment is used. This includes motorized raking machines and sometimes machines that act like large vacuum cleaners.

**FIGURE 8-8**
Once landscapes are installed, they must be maintained. This is a large industry. *Courtesy of Terri Hamlin, Georgia Department of Education.*

FIGURE 8-9
Plants sometimes have to be treated with chemicals to control pests. *Courtesy of Terri Hamlin, Georgia Department of Education.*

A general evaluation is then done. Pest damage is looked for as well as places where the lawn has died out. Sometimes a chemical treatment is needed if there is a serious pest problem. This may be done by licensed applicators. These people know the proper pesticide to use and the correct amounts. They also follow strict guidelines to prevent problems with the toxic materials.

Fertilizing is the next step. For grasses that grow well in the cool part of the year, the first fertilization is done in the early spring. For those grasses that grow best during warmer weather, late spring is a better time to fertilize. This is generally done in combination with a chemical weed killer.

For grasses to do their best they must have enough water. Usually water is added by using irrigation. This can be done with overhead irrigation, where sprinkler heads are used. Sometimes underground irrigation is used. This is where a water line is buried.

Irrigation is especially important in the early growing season. If grass does not get the water it requires when it first starts growing, it may remain stunted for the rest of the year.

By far the most time spent on a well maintained lawn will be mowing. If the grass is healthy and

FIGURE 8-10
Irrigation systems provide water for the landscape plants. *Courtesy of James Strawser, The University of Georgia.*

FIGURE 8-11
For grass to remain healthy and attractive, it has to be mowed at the proper height. *Courtesy of James Strawser, The University of Georgia.*

lush (like it should be) it will grow quickly. Many of you are probably very familiar by now with a lawn mower. No doubt some of you mow the lawn weekly on Saturdays. Many landscape professionals got their start mowing other people's lawns.

For grass to remain healthy and looking good, most need to be mowed at 2½ to 4 inches in height. This depends on the type of grass and the height the homeowner is comfortable with. In most cases such details are left for the maintenance person to decide.

With proper fertilization, pest control, irrigation, and mowing a lawn can usually be kept beautiful for many years.

## Pruning

**Pruning** is the removal of the branches of shrubs and trees. This is done to keep the size or shape of a plant the same. Sometimes part of the plant is damaged. The damaged portion of the plant is then pruned out. Afterward the plant may have to be pruned to a more pleasing shape.

Sometimes pruning is an art form. Many gardens in Europe have plants that have been pruned into

**FIGURE 8-12**
Pruning is the removal of some of the branches. This helps to shape the tree or shrub. *Courtesy of James Strawser, The University of Georgia.*

**topiary.** Topiaries can be anything from simple geometric forms to intricate animal shapes. Many painstaking hours go into these works of art.

In landscape maintenance, pruning is one of the most important procedures to do correctly. A plant that is pruned badly might never recover or it could actually die. The best times for plants to be pruned are usually spring or fall. This can sometimes interfere with a landscaper's busy seasons, however.

Plants that bloom in the spring should not be pruned in the fall. This cuts off buds that would

**FIGURE 8-13**
A topiary can be anything from a simple geometric form to an intricate animal shape. *Courtesy of Frank Flanders, The University of Georgia.*

normally bloom in the spring. Another problem with fall pruning is that pruning often stimulates a plant to produce new growth. This new growth is more susceptible to cold damage.

During the summer it is more difficult to actually see the branching habit of the plant. This makes it harder to decide correctly which branches to prune and which to leave. Also the plants are more susceptible to insect and disease damage at this time of the year.

Most professional pruning goes on in the winter. Not only does that allow the pruner to see exactly what he or she is pruning, it also provides work during an otherwise slow time. In addition, this is the time when many plants are in the dormant state. Less damage is done to the plant when it is dormant.

In the fall the landscape should be prepared for winter. Not only can low temperatures hurt plants, but so can the wind. During the active growing months most plants have leaves. These leaves help to deflect and slow down the wind. With the arrival of fall, many trees and shrubs drop their leaves. This allows the wind to blow directly on some plants that have been protected during the rest of the year.

**FIGURE 8-14**
Evergreens are sometimes planted as a windbreak. *Courtesy of James Strawser, The University of Georgia.*

In some instances, rows of evergreen shrubs and trees are planted as windbreaks. Windbreaks block the wind from houses, patios or more tender plants. In colder climates, the build up of snow or ice on the limbs of trees can damage them. Plants can be permanently bent or broken by the weight. Heavy loads of snow should be brushed or shaken off plants gently.

Winter damage can be caused by salt used on snow collected on walkways, drives and roads. The maintenance person must plan carefully to avoid this problem.

There are many aspects to the job of landscape maintenance. Most of the procedures involve forethought and planning as well as hard work.

# Interiorscaping

Think about a mall or other place of business you have been to recently. Most have some kind of **interiorscaping.** This is the concept of bringing the outside inside. Everything from a potted plant in the corner to large areas with beds of plants and fountains are parts of interiorscaping.

The interiorscaper faces problems that a landscaper does not. Plants do not grow naturally inside buildings. Most houses and public buildings were built with the comfort of human beings in mind, not the comfort of plants. Everything from light to water must be provided for interior plants. The media used to grow the plants must retain moisture and at the same time have good drainage.

The first step for developing a good interiorscape is the same as for a good landscape—planning. In landscaping, however, a designer can level a hill or use other drastic measures, use a wide variety of plants, or make large swaths of green healthy turf. An interiorscaper is much more limited. In

**FIGURE 8-15**
Interiorscapes bring the concepts of landscaping inside. *Courtesy of James Strawser, The University of Georgia.*

**FIGURE 8-16**
Interiorscapes require a lot of maintenance. *Courtesy of James Strawser, The University of Georgia.*

most instances the interiorscaper is left with a few areas that have been set aside for growing plants. Expanding or moving the planting beds is not an option.

Also, plants that stay healthy inside buildings are limited. Most must be tropical foliage plants. Fruit trees and most flowering shrubs cannot be grown inside because they must have a certain amount of cold before they will bloom.

Interior plants must be able to take a lot of abuse. Often there is pollution inside buildings that the plants must tolerate. Dust builds up on their leaves quickly. This reduces photosynthesis as well as making the plant look bad.

Usually there simply is not enough light inside for a plant to grow and remain healthy. In this instance, supplemental lighting must be provided. This means there must be a source of electricity as well as a place to anchor the lighting fixtures.

Plants must have water, whether they are inside or outside. This calls for more planning on the part of the interiorscaper. There must be a source of water nearby, and containers must be designed to hold water seepage. Clients may get upset if they have water stains on their carpets.

Indoor plants need fertilization just as their outdoor counterparts. They need less, however, and it should be applied less frequently.

Insects and other pests can damage indoor plants. It is especially important to spot these problems before they become serious. Because of the closed environment indoor plants grow in, pest problems can spread rapidly. There is usually a lot of foot traffic around indoor plants. Because of this, very careful use of pesticides is crucial.

With proper planning and maintenance indoor plantings can give a business or home a peaceful, comfortable atmosphere.

## Summary

Landscaping can be a well paid and fascinating occupation. There is more and more need for educated, dedicated people in this field.

Although it involves hard work, there is nothing like the feeling you get when you look at a beautiful landscaping job that you have planned, planted, and maintained.

# 8 CHAPTER REVIEW

## Student Learning Exercises

1. Closely examine the landscaping around your home or other building. Identify the principles of landscape design that were used.

2. Locate a building that is in need of landscaping. Draw a sketch of the building and design a landscape for it. Be sure to identify the principles of design you used.

3. Locate what you consider to be the most beautifully landscaped building in your area. Draw a sketch of the design and explain why you think the design is good. Share your opinions with the class.

# True/False

1. ___ All landscapes are balanced.

2. ___ Flower and plant beds should have smooth, rounded shapes.

3. ___ Unity is usually the hardest principle to master.

4. ___ Lawn maintenance begins in the winter and continues until early spring.

5. ___ Irrigation is especially important early in the growing season.

6. ___ The most time spent on a well maintained lawn will be for mowing.

7. ___ A plant can die from being pruned wrong.

8. ___ During the summer it is easy to see the branching habit of a plant.

9. ___ Indoor plants need less fertilization than outdoor plants.

10. ___ Interior plants must be able to take a lot of abuse.

# Multiple Choice

1. A person who makes landscape designs professionally is called a
   a. landscape architect.
   b. professional landscaper.
   c. land maintenance manager.

2. The removal of the branches of shrubs and trees is called
    a. topiary.
    b. pinching.
    c. pruning.

3. Scale means that all plants should be
    a. proportional to each other.
    b. the same size.
    c. different sizes.

4. A type of formal placing of plants is called
    a. scale.
    b. balance.
    c. unity.

5. Limit flower borders to two or three basic
    a. sizes.
    b. shapes.
    c. colors.

6. The principle that is usually hardest to master is
    a. simplicity.
    b. balance.
    c. unity.

7. The most time and money in the landscape maintenance business is spent on
    a. interiorscaping.
    b. lawn care.
    c. irrigation.

8. Most professional pruning occurs in the
    a. fall.
    b. spring.
    c. winter.

9. Rows of tall evergreen shrubs and trees are planted as
    a. wind breaks.
    b. design trees.
    c. topiary.

10. Less damage is done to the plant when it is
    a. growing.
    b. dormant.
    c. fertilized.

# Discussion

1. Explain what is meant by landscaping.

2. What are the five basic landscaping design principles?

3. How is unity expressed?

4. What are some problems with indoor plants?

5. What is a landscaping plan?

6. What is meant by landscape maintenance?

7. When is the best time to fertilize?

8. Why is irrigation important early in the growing season?

9. What four things are most important in keeping a lawn beautiful?

10. Why are leaves important to plants?

# 9

# Fruit and Nut Production

## Student Objectives

*When you have finished studying this chapter, you should be able to:*

- Explain how consumers can buy fresh fruit all year long.
- List the major fruit and nut crops in the United States.
- Describe how pome fruits are produced.
- Describe how Stone fruits are produced.
- Describe the citrus industry.
- Discuss nut production in the United States.

### KEY TERMS

pome fruits
grafting
cultivars
stone fruits
citrus fruits

I f you go into any large grocery store you can find a wide variety of fruit. Fresh fruit of many types is readily available. Apples, peaches, cherries, oranges, strawberries, pears, bananas, and many other fruits can be bought at any time of the year. This has not always been the case. Fresh fruit spoils very easily. Until relatively recently, people could only eat fresh fruit during the time between when the fruit ripened and when it spoiled. Fruit was preserved, but preserved fruit was never the same as fresh.

**FIGURE 9-1**
Many types of fresh fruit can be bought all year long. *Courtesy of James Strawser, The University of Georgia.*

**FIGURE 9-2**
Much fruit is still preserved. *Courtesy of James Strawser, The University of Georgia.*

The fruit was dried, canned, or made into jelly and preserves. Even though a lot of fruit is eaten fresh, much fruit is still preserved by these methods.

Modern transportation and storage techniques allow us to move fresh fruits long distances. In the southern hemisphere the seasons are opposite from ours. While it is winter in North America, it is summer in that part of South America that is south of the equator. When our fruit trees are dormant, theirs are in full production. Fruit is shipped from South America and we can have fresh fruit in the middle of the winter. Peaches, pears, plums, and grapes are shipped to the United States during the winter. Most of the fresh apples we enjoy have been stored from the U.S. crop.

## Pome Fruits

**Pome fruits** are those that have fleshy insides with small seeds. Only two pome fruits of importance are grown in this country—apples and pears. Most of the apples are grown in Washington, New York, Michigan, California, and the higher altitudes of the

Appalachians. Apples grow so well in these areas because the trees need a cool climate. Cool nights during the ripening process make the apples have a good flavor and color. In the South the nights are warmer during the time when the fruit ripens and the quality of the fruit is not as good.

About half of the apples produced are marketed fresh. The rest are processed into juice, jellies, pies, and other products. The most popular varieties are Red Delicious and Yellow Delicious, which are sold as fresh fruit. Other varieties include Macintosh, Rome Beauty, and Jonathan. These varieties are usually used for processing or cooking.

Most of the pears in this country are grown in Washington, Oregon, and California. Pears do well in the semiarid regions of these states. Pears are more difficult to produce in the East because of disease. A disease called fire blight attacks pear trees in the more humid areas. This blight is difficult to control and makes the pears expensive to produce. Many of the same practices used in the growing and managing of apples are used in growing pears.

**FIGURE 9-3**
Apples do best in climates where the nights are cool during ripening. *Courtesy of James Strawser, The University of Georgia.*

**FIGURE 9-4**
Most pears are grown in California, Oregon, and Washington. *Courtesy of Washington State University.*

**FIGURE 9-5**
Pruning helps produce a more desirable tree. *Courtesy of James Strawser, The University of Georgia.*

**FIGURE 9-6**
Through grafting, several varieties can be grown on the same tree. *Courtesy of USDA.*

## Apple Production

The first step in the production of apples is the production of young trees. Instead of growing the trees from seed like many other plants, a process called **grafting** is used. Apples are difficult to propagate by any other method. For example, a hybrid apple tree might be difficult to grow from seed because the tree will be different from either of its parents. If a young sprout is combined with a portion of a desirable tree, the results can be a tree that grows and produces like the desirable tree. This method is also used to produce specialty trees such as dwarf trees and fruit trees that will grow several different varieties of fruit. Trees are also grafted onto rootstock that is stronger or more suited to the area.

Grafting involves cutting two parts of a tree. The lower part is called a rootstock and the upper part is called a scion (sometimes spelled "cion"). The rootstock may be larger than the scion. This gives the advantage of faster growth and earlier maturity. The two cuttings are carefully aligned and wrapped in string. The joint is then coated with wax and the plant is set in the soil. Eventually the two plant parts grow together and form a new tree. Through grafting, several varieties of apples can be grown on the same tree. Scions from different varieties are grafted on to different branches of a tree. Each grafted limb will grow the same type fruit as the tree from which the scion grew.

The young trees are set out in evenly spaced rows. Setting out in rows provides room for mowing and spraying machinery to maneuver. In arid areas, irrigation systems may be installed. The young trees are kept free of weeds and sprayed to prevent damage from insects and disease. Also, the young trees are pruned to shape them and to prevent the growth of weak branches.

It generally takes several years for the trees to begin bearing fruit. The length of time varies according to the variety, climate, soil conditions, management practices, and size of the trees. Because of the length of time involved before the trees produce, producers have a large investment in the trees.

Some trees, especially certain **cultivars,** are self-sterile. Cultivars are plants that are cultivated and grown by humans. These plants may have been bred up to the point where they are sterile unless pollinated from other plants. Red Delicious apple trees are a good example. By themselves, they are sterile and need another apple tree to cross-pollinate. Yellow Delicious trees produce pollen that is excellent in fertilizing Red Delicious trees.

Fruit trees are generally pollinated by insects. Remember from Chapter 4 that honeybees are particularly adept at pollinating.

**FIGURE 9-7**
An orchard represents a sizable investment before fruit is produced. *Courtesy of James Strawser, The University of Georgia.*

**FIGURE 9-8**
Most fruit trees have to be pollinated by bees or other insects. *Courtesy of James Strawser, The University of Georgia.*

Fruit growers use bees to pollinate the trees. Hives are set in the orchards every year during the spring. The bees work the blossoms and also produce honey. Honey from fruit blossoms is delicious and is in great demand.

Once the trees have bloomed and the fruit has set, the trees have to be sprayed regularly to prevent insect and disease damage. Both disease and insect damage lowers the quality of the fruit. Modern insecticides are safe when used correctly. The USDA carefully monitors the pesticide residues on fruit.

When the apples are ripe, harvesting begins. Most apples are harvested by hand to make sure the fruit is not bruised. Bruises lower both the appearance and flavor of the fruit. Pickers are careful how they handle the fruit. The fruit is gathered in small containers that do not damage the fruit.

Apples that are to be stored as fresh fruit receive special handling at the processing plant. Once the fruits are washed and cleaned, they are coated with a thin coating of paraffin to prevent damage to the skin and to help seal the surface of the fruit. The ap-

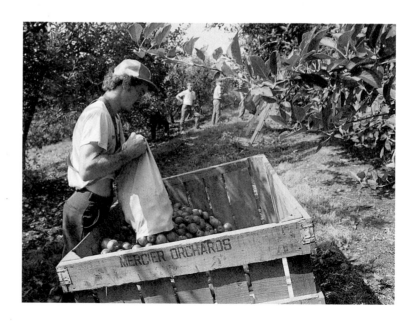

**FIGURE 9-9**
Pickers have to be careful not to damage the fruit. *Courtesy of James Strawser, The University of Georgia.*

ples are stored in a process called controlled atmosphere storage. The fruit is placed in dark chambers where the temperature is held at a constant temperature of between 32 and 55 °F. Oxygen and carbon dioxide levels are decreased and are replaced by nitrogen. The apples are maintained in a proper balance of oxygen, carbon dioxide, and nitrogen. These gases, combined with the correct cool temperature and the absence of light, allow apples to be kept fresh for many months.

**FIGURE 9-10**
Properly stored, apples can be kept fresh for a year or more. *Courtesy of Henry Waelti, Washington State University.*

## Stone Fruits

**Stone fruits** have large seeds called pits or stones in the center. The most commonly grown stone fruits are peaches, cherries, plums, and apricots. All of these fruits are grown worldwide and in the United States.

### *Peach Production*

Peaches are grown all across the country. The major states are California, South Carolina, Georgia, New Jersey, Pennsylvania, and Michigan. Peaches need a certain number of hours of cold weather to bring the trees through the process of dormancy. Most varieties need around 750 hours of weather below 45 °F in order to have the proper amount of blossoms and fruit.

There are two types of peaches. One type is known as freestone. These varieties of peaches have stones that easily separate from the flesh when the peach is cut in half. The other type, clingstone, does not separate from the flesh of the peach. Most of the peaches that are marketed fresh are freestone and most clingstone are used for processing.

Like most fruit trees, peaches are grafted onto rootstock. They are set out in rows and managed by pruning, fertilizing, and spraying. Peach trees are

**FIGURE 9-11**
Peaches are grown all across the country. Different varieties grow in different areas. *Courtesy of James Strawser, The University of Georgia.*

pruned differently than apples. Peach trees are pruned to have an open, inverted umbrella shape. This is to allow more sunlight into the branches of the tree.

Most peaches are harvested by hand. The fruit is so soft that it bruises very easily. Fruit that is intended for the fresh market must be free of bruises and blemishes. Peaches that are intended for pro-

## Career Development Events

Everyone likes to receive recognition. The FFA Proficiency Awards and Supervised Agricultural Experience Program (SAEP) are two ways every student can participate and become winners. To begin you must first become an FFA member and be enrolled in an agriculture class. Your teacher will help you design an SAEP that will put into practice your interests plus the skills and abilities learned in the classroom, shop, and laboratory. By becoming active in an SAEP, you will learn how to earn money that you may reinvest, spend, or save. You will also develop the skills required to produce, operate, and market your selected product. The Fruit and Vegetable Production Proficiency Award recognizes individual FFA members who develop and manage

The Fruit and Vegetable Production Proficiency Award recognizes individual FFA members who excel in producing and marketing fruits or vegetables. *Courtesy of Blane Marable, Morgan Co. High School, Madison, GA.*

cessing may be harvested by a mechanical picker. These machines have a flexible platform that goes around the tree. The machine grabs the tree and gently shakes it. Ripe fruit falls onto the platform and the peaches are gathered in. The fruit is rushed to market where it is washed, peeled, and processed. The processing may include canning, pie production, or making into preserves.

SAEPs in producing and marketing products such as field grown vegetables, orchard fruit, and vineyard crops. Recordkeeping is a very important part of a Supervised Agricultural Experience Program (SAEP). The student must maintain a list of fruit and/or vegetable production skills that are learned in the classroom and/or used during the SAEP. Development of equipment and material inventories plus the recording of income, expenses, and accomplishments must be accurately kept. The student must make managerial decisions involving the production and marketing of the fruit and/or vegetables and also provide the labor in planting, maintaining, and harvesting. An award-winning proficiency is the result of a long-term SAEP, good recordkeeping and a student's involvement with the local FFA and community.

Kerry won the Fruit and Vegetable proficiency award in his FFA chapter. He had grown a small vegetable garden his first year in agriculture and won the chapter garden contest. In his second year he rented 1 acre of land and raised tomatoes and squash and sold them at the local farmer's market. Kerry added the production of cool season crops (lettuce and cabbage) his third year of agriculture. After the fourth year of his SAEP, Kerry had made enough income to purchase 2 acres, where he plans to continue his vegetable business. Kerry was on his high school football team and chaired the FFA Food for America program for two years. While attending the state FFA convention, Kerry placed second in the tractor driving contest. Kerry's experiences with the FFA and his SAEP have prepared him to be a successful leader, businessperson, and citizen.

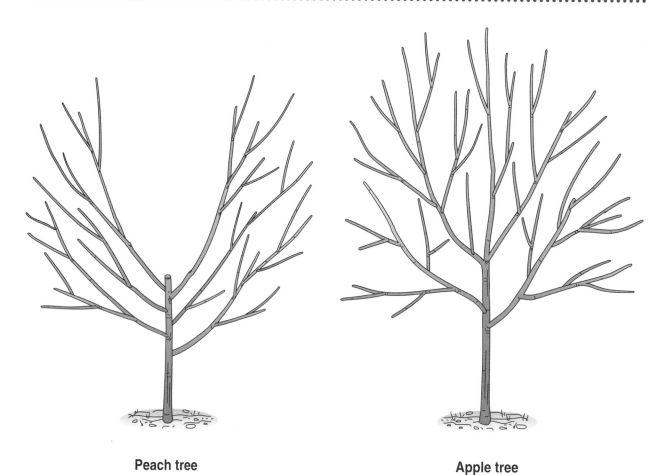

**Peach tree**                                    **Apple tree**

FIGURE 9-12   Peach trees are pruned to have an open center.

## Citrus Fruit

**Citrus fruits** have been grown for thousands of years. They are thought to have originated in Southeast Asia. They were introduced to Europe in the 12th century and were brought to America by the European settlers. These fruits provide us with a good tasting, nutritious addition to our diets. Citrus is an important source of vitamins and minerals, particularly vitamin C. During the 1750s a British doctor named James Lind noticed that sailors who spent long tours at sea frequently came down with a disease called scurvy. The sailors were fed on a diet of dried beans and dried meat. Dr. Lind discov-

**FIGURE 9-13**
Citrus is grown in the extreme southern areas of the United States. *Courtesy of Vick Christ.*

ered that by giving the sailors citrus fruit, scurvy could be avoided. Later it became known that the vitamin C in the fruit prevented the disease.

The citrus grown in the United States includes oranges, grapefruit, tangerines, tangelos, lemons, and limes. Most of the citrus is grown in Florida, California, Texas, and Arizona. These areas are about the only places in the United States with a climate warm enough to produce the fruit.

Although much of the crop is sold as fresh fruit, there is a large industry that processes citrus. The largest processed products are orange, grapefruit, and lemon juices. The fruit is crushed to remove the juice. Much of the water is removed from the juice, and it is packaged and frozen. The consumer can then thaw the concentrate, add the proper amount of water, and enjoy orange juice.

One interesting by-product of the orange and grapefruit juice industry is the pulp. Pulp is the fibrous material left over when the juice is pressed out of the fruit. This material makes a valuable cattle feed. When mixed with grain, citrus pulp is used to fatten cattle for market. A good use has been found for a material that once went to waste.

**FIGURE 9-14**
Citrus juice is concentrated, then frozen. *Courtesy of James Strawser, The University of Georgia.*

## Nut Production

Nuts are actually seed kernels that are housed in a woody structure. They have been used as food for as long as people have been on the earth. Most nuts are high in proteins and fats. Nuts may contain as much as 600 to 700 calories per 100 grams compared to 250 calories per 100 grams for beef with moderate fat content.

Each year producers in the United States harvest over 430,000 tons of nuts. The nuts are grown in the warmer regions of the country. While nut trees will grow in all states, they are generally sensitive to the cold and grow better in the South. Four types of nuts are of real commercial importance in the United States: almonds, pecans, walnuts, and filberts.

More almonds are produced in the United States than any other nut. Almost all of the almonds grown in this country are produced in California. The almond is native to the hot, arid regions of West Asia and California matches the climate. The almond tree resembles the peach tree. In fact, peach trees are often used as root stock for grafting almond trees.

Although there are over 100 varieties of almonds grown, there are two major types of almonds. The

**FIGURE 9-15**
Almost all almonds grown in the United States are grown in California. *Courtesy of Jack Kelly Clark, University of California, Statewide IPM Project.*

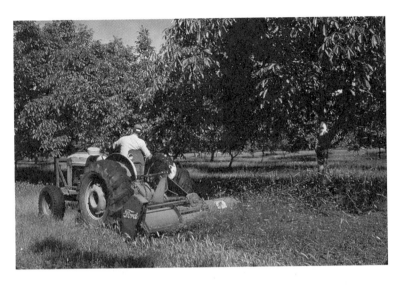

**FIGURE 9-16**
English walnuts are grown in California and Oregon. *Courtesy of Jack Kelly Clark, University of California, Statewide IPM Project.*

sweet almond is edible and is grown for food. The bitter almond is grown for the oil in the kernel. The sweet, edible almond is eaten raw, roasted, or in candies and pastries. The oil from the bitter almond is used to produce prussic acid that is valuable as a fumigant. The oil is also used in some medicines.

The major type of walnut is the English walnut that is grown in California and Oregon. The black walnut is also grown, but it has smaller kernels that are more difficult to remove from the hull. Most English walnuts are grafted onto black walnut rootstock because the black walnut trees have stronger root systems.

Most walnuts are sold in the shell although some are sold as kernels. These shelled nuts are used for baking and in salads. Another use for the walnut trees is the valuable wood. Walnut has always been a highly prized furniture wood.

Pecans are classified as genus *Carya*, which includes the hickories. Even though this nut tree is classified as a hickory, the nuts differ from other hickory nuts. Pecans have larger kernels and thinner shells. In fact, some pecan varieties are known as "paper shell pecans" because of the very thin shells.

FIGURE 9-17
Pecan trees grow quite large and mature slowly. *Courtesy of James Strawser, The University of Georgia.*

FIGURE 9-18
Nuts are removed from the tree by a machine that gently shakes the tree. *Courtesy of James Strawser, The University of Georgia.*

Pecans are grown in the South. The leading states are Georgia, Alabama, Texas, Louisiana, New Mexico, and Oklahoma. The trees grow quite large (75 to 90 feet) and mature slowly. Most need eight to ten years from the time they are set out until they begin to produce. The trees live for many years and produce a long time.

Pecans are sold both shelled and unshelled. The nuts are used in pastries, pies, and other sweets. They are relatively high in fat content and contain a considerable amount of protein.

Filberts are also known as hazelnuts. Most are grown in Oregon and Washington. They are eaten raw or roasted and are becoming increasingly popular. One of the newer marketing techniques is to make consumers aware of the fact that filberts are hazelnuts. Consumers generally do not recognize the name filberts. The nuts grow on either a small tree or shrub depending on the management practices used. The trees bloom in the winter and the nuts mature in the fall.

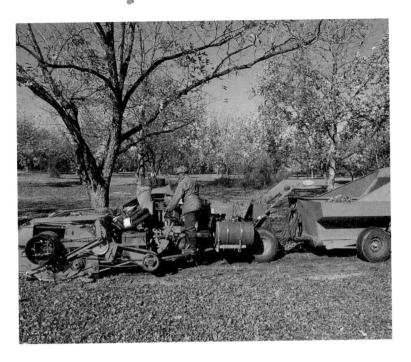

**FIGURE 9-19**
The nuts are picked up from the ground by a nut sweeper. *Courtesy of James Strawser, The University of Georgia.*

Nuts are harvested by using a tree shaker. The machine grabs the tree and gently shakes it until the nuts fall off. Another machine sweeps the nuts into rows called windrows. A large vacuum picks the nuts off the ground. This means that the ground beneath the trees needs to be kept clean of weeds and foreign objects such as rocks. At the processing plant the nuts are shaken to remove rocks and limbs. They are then placed in large bins where the excess moisture is removed.

## Summary

The fruit and nut industry in the United States is very large. This industry provides consumers with food that is delicious and nutritious. Our lives are made better through the availability of fruits and nuts all year round. In addition they greatly add to our health.

# 9 CHAPTER REVIEW

## Student Learning Activities

1. Visit a large grocery store and make a list of all the different types of fresh fruit that are available. Determine what country produced the fruits. Compare your findings with others in the class.

2. With your teacher's assistance, slice up several different varieties of apples and ask people in the class to taste the samples. Keep a list of the ones favored by the most people. Ask each person to tell why he or she preferred a particular variety.

3. Survey your area and determine what types of fruit and nuts are grown in your area. Report on the climatic conditions and other factors that make your area good for producing these crops.

## True/False

1. ___ Fresh fruit spoils very easily.

2. ___ Most of the fresh apples we enjoy come from South America.

3. ___ Apple trees do best in cool climates.

4. ___ Fruit trees are generally pollinated by insects.

5. ___ Peaches, cherries, plums, and apricots are examples of stone fruits.

6. ___ Citrus pulp makes a good cattle feed.

7. ___ Nuts are grown in the cooler regions of the country.

8. ___ More pecans are produced in the United States than any other nut.

9. ___ Hazelnuts are also known as filberts.

10. ___ Nuts are harvested by using a tree shaker.

# Multiple Choice

1. The only two pome fruits of importance that are grown in this country are apples and
   a. oranges.
   b. pears.
   c. cherries.

2. The most popular variety of apple is
   a. Granny Smith.
   b. jonathon.
   c. red delicious.

3. Fruit growers make use of bees to pollinate the
   a. flowers.
   b. seeds.
   c. roots.

4. The two types of peaches are freestone and
   a. filberts.
   b. sandstone.
   c. clingstone.

5. Most of the peaches that are marketed fresh are
   a. freestone.
   b. clingstone.
   c. sandstone.

6. Citrus fruits are thought to have originated in
   a. North America.
   b. South Africa.
   c. Southeast Asia.

7. Most nuts are high in fats and
   a. filberts.
   b. proteins.
   c. seeds.

8. Almost all of the almonds grown in the United States are produced in
   a. Washington.
   b. Georgia.
   c. California.

9. The almond tree resembles the
   a. peach tree.
   b. apple tree.
   c. walnut tree.

10. Pecans are grown in the
    a. West.
    b. North.
    c. South.

# Discussion

1. In the past, people could only eat fresh fruit during a very brief time of the year. Why?

2. What are some ways fruit is used?

3. Why are cool nights beneficial to apples?

4. What is fire blight?

5. What are the two types of peaches?

6. Why are peach trees pruned into an inverted umbrella shape?

7. How was the importance of vitamin C in citrus fruits discovered?

8. In the United States, where is most citrus fruit grown?

9. What three citrus fruits are processed most often?

10. What are the two types of almonds?

# 10

# Row Crops

## Student Objectives

**When you have finished studying this chapter, you should be able to:**

- List the major row crops grown in the United States.
- List the types of grain that are produced in the United States.
- Describe the production and uses of grain crops.
- List the types of oil seed crops that are produced in the United States.
- Describe the production and uses of oil seed crops.
- List the types of fiber crops that are produced in the United States.
- Describe the production and uses of fiber crops.

### KEY TERMS

row crops
grain
flour
cereal
oil crops
fiber crops

A large portion of our gigantic agricultural industry is row cropping. **Row crops** are those that are planted and cultivated in rows. Usually it refers to annual plants that have to be replanted each year. Every state in the country grows row crops. The type varies with the climate, soil type, and available markets. Some crops, like corn, are grown in all 50 states.

Hundreds of different types of crops are grown in this country. They may be classified as grains, oil crops, and fiber crops. It would be impossible to describe all of the different types of crops grown in this country. However, a few major crops comprise most of the row crops in this country. These major crops will be described in this chapter.

## Grain Crops

Probably the very first crop that people planted was **grain.** Grain is the seed from plants that are classified as grasses. Humans discovered that not only were the grass seeds good to eat, but they could be planted and more grass that produced more seed would grow. The grasses that produced the most and best seed were selected for cultivation. As more was grown year after year, plants were developed that produced more seed that could be used for food.

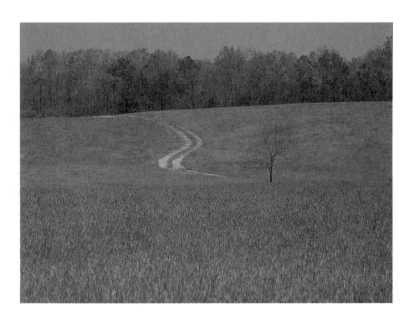

FIGURE 10-1
Grasses that produce the most and best seed are cultivated as grain.
*Courtesy of James Strawser, The University of Georgia.*

## Wheat

Wheat is perhaps the oldest of the major grain crops. Ancient religious documents mention the growing of wheat. Most civilizations have grown wheat to be used in the making of bread. Bread is a staple food that can be preserved and provides nutrition.

Most of the wheat grown is used for human food. It has been the basis of bread for thousands of years. The seed kernels are ground into **flour,** and flour is used in baking. Wheat flour contains a high percentage of a substance called gluten that causes the flour particles to stick together when moisture is added. This sticky dough traps gases from fermentation that causes the bread to rise. When compared to other grains, wheat flour is relatively high in protein and makes a good tasting food.

Wheat is grown all across the country, but most of the production is centered in the Midwest and Great Plains. Several types of wheat are grown in this country. These include hard red winter, soft red winter, hard red spring, white, and durum wheat.

**FIGURE 10-2**

A substance in wheat flour called gluten makes it good for baking bread. *Courtesy of James Strawser, The University of Georgia.*

**FIGURE 10-3**
Hard red spring wheat makes the best flour for bread. *Courtesy of John Deere & Company.*

Hard red winter wheat is grown in Kansas and nearby states. More hard red winter wheat is grown than any other type. It contains a relatively high percentage of gluten that causes the flour to be among the best for baking breads. It is second only to hard red spring wheat for baking breads.

Soft red winter wheat is grown in areas of the Midwest where there is abundant rainfall. The moisture causes the wheat grains to be softer, more starchy and lower in protein than the hard red winter wheat. Consequently gluten content is lower and the baking quality of the flour is not as good.

Hard red spring wheat has the highest amount of gluten of the commonly grown wheats. The highest quality breads are made from hard red spring wheat. This wheat is grown in the upper Great Plains area of North and South Dakota, Montana, and Minnesota. The weather gets too cold in these areas to grow winter wheat, so much of the wheat production is spring wheat.

White wheat is grown in the states of Washington and Oregon with some being grown in Michigan and New York. White wheat is grown both as winter and spring wheat. The flour from white wheat does not make very good bread and is used mostly for baking crackers, pastries, and breakfast **cereal** foods.

Durum wheat is grown through the upper Great Plains. It is more drought- and disease-resistant than the hard red winter wheat. Durum wheat contains a high percent of gluten but it is considerably lower quality than the hard red winter and spring wheats. Durum is used for making pastas like macaroni and spaghetti.

Winter wheat is planted in the fall and harvested in mid to late summer. Spring wheat is planted in the early spring and harvested in the late summer and early fall. At planting, the seeds are placed in the ground using an implement called a grain drill. The drill places seeds in close rows (6 to 12 inches apart) at the proper depth in the soil. Winter wheat must go through a period of prolonged cold before it can form grain heads. This process if called vernalization.

Wheat harvesting is done with a combine. This machine cuts the wheat stalks off and threshes the grains of wheat out of the seed heads. A large fan blower separates the hulls and chaff from the grain. The grain is then conveyed to a bin that when full is emptied into a truck. The truck takes the grain to

**FIGURE 10-4**
Wheat is planted using an implement called a grain drill. *Courtesy of Dewey Lee, The University of Georgia.*

**FIGURE 10-5**
Grain is harvested using a combine. *Courtesy of James Strawser, The University of Georgia.*

an elevator where it is sampled, graded, and stored. The producer is paid on the basis of the grade of the wheat. The grade is based on moisture content and the amount of weed seed and other foreign matter in the grain.

## Corn

More corn is produced in the United States than any other crop. The United States produces almost half of all the corn produced in the world. It is one of the few major crops that is native to this country. Prior to the European settlers arriving in the new world, corn was not grown in other parts of the world. However, it was widely cultivated by Native Americans in both North and South America.

Corn is grown in every state in the country, but the majority of the production is in the Midwest. In this region the soil is deep and rich and the rainfall is adequate. This combination allows for huge yields of corn. These yields have dramatically increased over the past few decades. Research into methods of production has greatly increased production. However, the greatest impact has come from hybrid varieties. A hybrid is a cross between two different varieties. The resulting plant outproduces either of the parents. In 1930, when very little hybrid corn was grown, the average production was around 25 bushels per acre. Today, when almost all corn is hybrid varieties, the average yield is almost 150 bushels per acre. This average is for all of the country. A well-managed field in the Midwest may average well over 200 bushels per acre.

Corn is used both for human food and animal feed. Almost all of the corn produced is either white or yellow. Of the two types, yellow corn is by far the most widely grown. White corn is grown for corn meal that is used in baking and bread making. It is also used in making breakfast cereals and corn

FIGURE 10-6

Corn is grown in every state in the United States. *Courtesy of James Strawser, The University of Georgia.*

FIGURE 10-7
A well-managed field of corn in the Midwest may average as much as 200 bushels of corn per acre. *Courtesy of John Deere & Company.*

chips because it toasts better than yellow corn. Yellow corn contains a substance called carotenoid pigments that animals can convert to vitamin A. Both yellow and white corn are used for animal feed because they are so high in carbohydrates. Cattle and hogs are fattened on corn because it is so high in this nutrient. Because corn is relatively low in protein, a protein supplement is usually added to livestock feed.

Corn has other uses. Cooking oil is pressed from the grain. It has no cholesterol and is of very high

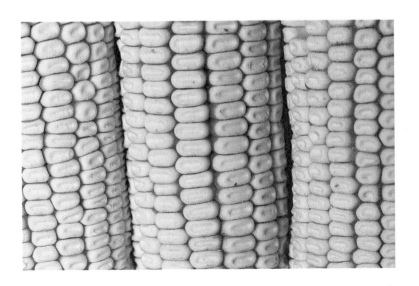

FIGURE 10-8
White corn is used more for human food. Yellow corn is used more for animal feed. *Courtesy of James Strawser, The University of Georgia.*

quality. Corn starch is produced from the starchy content of the grain. It is used in thickening agents in puddings and pies, as a drying agent in body powder, and a broad array of other uses.

A major new use of corn is the production of ethanol alcohol. Alcohol made from corn and other

## Career Development Events

FFA members have a great opportunity to compete for awards in the area of row crops, including grain, oil, and fiber products. These awards encourage the use of new technology and management practices in the production and marketing of agriculture crops. By taking agriculture classes, the student can learn and practice the knowledge and skills required to become an effective producer of agriculture products. Producing crops is a great Supervised Agricultural Experience Program (SAEP) for any FFA member. The Agriculture teacher, student, and guardian should agree on the scope and focus of any SAEP. Students will be responsible for making managerial decisions about the ground preparation, planting, maintenance, harvesting, and marketing of the crop. The student must keep a list of crop production skills used in the classroom and in the field. Income, expenses, and yields of the crops produced and

harvested must also be accurately recorded. The building of inventory such as land, seed, fertilizer, machinery, and harvested crops over several years is essential to an award-winning proficiency.

Jill is an active member of her FFA. In her senior year she won the state agriculture management contest and a scholarship to her state's agriculture college. Jill began her SAEP in the ninth grade. She lived on a 100-acre family farm, and part of the signed agreement between herself, her father, and agriculture teacher was to lease 2 acres of land to Jill. The first two years Jill tilled, sowed, managed, and harvested cotton on the 2 acres. Jill's agribusiness class taught her record keeping skills and investment opportunities that helped her reinvest her income wisely. Jill chose to sell her cotton with a local co-op. The last two years Jill diversified her production by crop rotations. After paying all the expenses of her operations, Jill's profit margin was small, but the experience was priceless. Jill's application for grains can be made to run car, truck, tractor, and other engines as efficiently as gasoline with much less pollution of the environment. The technology for this process is hundreds of years old and is really a simple procedure. Alcohol produced from corn holds promise for replacing much of the

the crop proficiency award won chapter, district, and state recognition. Jill is one of many who have found the FFA to be the foundation of her future.

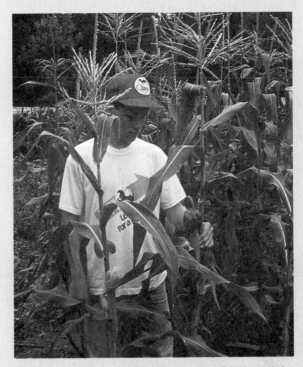

Crop production has been a part of FFA activities for years. Students can still receive awards in this area. *Courtesy of Blane Marable, Morgan Co. High School, Madison, GA.*

Frank is a high school student who lives in the city. This year he is applying for the chapter placement in agricultural production proficiency. Frank has been employed on a forage production farm for the last three years. Frank's FFA membership started in the seventh grade. He continued to be an active member and competed on the forestry and land judging teams. His sophomore year Frank came to his teacher looking for an after-school job. Frank's desire to be outdoors and his accomplishments in agriculture mechanics class led his teacher to place him on a local forage production farm just outside the city. Frank has kept accurate records of his hours of work, job skills, income, and expenses related to his job. Through the years Frank was promoted from field hand to large equipment operator. His experiences and skills have prepared Frank for a career in agriculture, and he is an excellent candidate for the proficiency placement award in agricultural production.

**FIGURE 10-9**
Some engines make use of ethanol that is made from corn. *Courtesy of Illinois Research Agricultural Experiment Station.*

petroleum-based fuels used today. At our high rate of usage, petroleum oil will someday run out. A mixture of ethanol and gasoline called gasohol now replaces over 40 million barrels of crude oil each year.

## Oil Crops

A major part of the row crops in this country is produced for their oil content. The two most widely produced **oil crops** are soybeans, and peanuts. The

**FIGURE 10-10**
Oil from the seed of plants is used for cooking oil. *Courtesy of James Strawser, The University of Georgia.*

oil from the seeds of these plants is used in a variety of ways ranging from cooking oils to paints to a replacement for diesel fuel.

## Soybeans

Soybeans probably originated in China, where they have been grown for thousands of years. They were imported into this country in the late 1800s. Through the development of varieties suited to the climate of various parts of the United States, soybeans have became a major crop in this country. In fact, the United States produces over half of the world's supply of soybeans. It is interesting that the United States now exports soybeans to China, the country in which the crop originated. This crop is grown extensively throughout the South and Midwest.

One reason that soybeans are popular is the wide variety of uses. One of the primary uses is for oil. The oil is pressed out and used for cooking and making margarine, shortening, cheese spreads, and coffee whiteners. New uses for the oil, such as a substitute for diesel fuel and as the basis for a high quality ink, have been discovered. The whole bean can be used to make meat substitutes and extenders.

The material left when the oil is pressed out is used for animal feed. The pressed cake is ground into a meal that contains an average of 45 percent protein. This high-quality feed is mixed with corn and other grains and fed to pigs and chickens.

## Peanuts

The peanut is native to South America and was transported to Africa by Europeans. From here it was brought to North America by Africans who were brought in as slaves. This nutritious food gained popularity in the latter part of the 1800s when the

**FIGURE 10-11**
The United States produces over half the world's supply of soybeans. *Courtesy of James Strawser, The University of Georgia.*

**FIGURE 10-12**
Soybeans are used for a variety of purposes ranging from animal feed to printer's ink. *Courtesy of James Strawser, The University of Georgia.*

FIGURE 10-13
Peanuts grow on the roots of vines that run along the ground. *Courtesy of James Strawser, The University of Georgia.*

boll weevil destroyed much of the cotton crop of the South. Scientists like George Washington Carver invented new uses for the peanut. Products like peanut butter became popular and a new industry grew up in the South.

Today, this crop is grown in many parts of the South. The leading states are Georgia, Alabama, Texas, and North Carolina. The nuts grow on the roots of vines that grow along the ground. They are planted in rows about 3 feet apart. At time of harvest the vines may completely cover the ground. The plants require a long growing season with warm weather. In addition, since the nuts grow in the ground, they must be grown in light sandy soil. Peanuts grown in soils with heavy clays are difficult to harvest.

At harvest, a large plow turns the roots toward the sky. The vines are left in the field for several days until the nuts begin to dry. A large machine picks up the vines and removes the nuts. The nuts are allowed to dry and are stored or processed.

Peanuts are sold in the shell raw, shelled raw, roasted in the shell, shelled and roasted, or boiled. Products such as peanut brittle, a variety of candies, and pastries are made from peanuts. Of course all of

FIGURE 10-14
A large plow turns the peanuts toward the sky. *Courtesy of James Strawser, The University of Georgia.*

us enjoy eating peanut butter that is made from ground up roasted peanuts.

One of the most valuable products from peanuts is peanut oil. This is one of the highest quality cooking oils. Like soybeans, peanuts are pressed by machines until the oil is removed. The cake that is left is used as a high-quality livestock feed. Not only is it high in protein content (around 50 percent), but the protein is very high quality.

## Fiber Crops

Much of the cloth produced today (such as rayon, nylon, and polyester) is manufactured from petroleum. In a process that was developed in the first half of the 20th century, petroleum is processed into long fibers that are used to make cloth. This process gives us cloth that is relatively inexpensive and very durable. For these reasons, **fiber crops** declined in popularity for several years. However, today, natural fibers are rebounding as a clothing material. Human-made fibers may be more wrinkle resistant and durable but they cannot match fiber produced by agriculture for comfort. Also, a lot of

**FIGURE 10-15**
Human-made fibers cannot match cotton clothing for comfort. *Courtesy of James Strawser, The University of Georgia.*

cloth produced is a combination (called a blend) of artificial and natural fibers that gives the flexibility and durability of synthetics and the comfort of natural fibers. The use of natural fibers makes us less dependent on imported petroleum.

Two main kinds of natural plant fibers produced for cloth are grown in the United States. These are cotton and flax, and both of these are also used as oil crops.

## Cotton

Cotton was probably first used in the Nile Valley in ancient Egypt. The climate and the fertile soil were ideal for the growth of the plant. In North America, humans have used cotton for at least 7,000 years and its cultivation goes back hundreds of years before the European settlers arrived. Much of the history of the United States developed around the growing of cotton.

Cotton is a shrub-like perennial plant (although it is cultivated as an annual plant) that requires a long growing season and warm temperatures. This

**FIGURE 10-16**
Cotton requires a long growing season. *Courtesy of James Strawser, The University of Georgia.*

means that the crop must be grown in the southern parts of this country. The leading states are California, Texas, Arizona, and the states of the lower Southeast.

About mid season the cotton flower blooms from a bud that appears on the stem. As it opens, the flower is a pale yellow that gradually turns pink and then red. After the flower is fertilized, the seed pod, known as a boll, forms. This structure is shaped like a small football and contains the fiber and the seed. The seed grows within the boll and the fibers form on the seeds.

Cotton is picked by large machines that use rotating finger-like projections on opposite rotating drums to remove the lint from the bolls. This ingenious invention replaced long strenuous hours of manual labor that was once required to handpick the cotton. The lint is removed from the picker and is either stored in the field in long loaf-shaped modules that hold 10 to 12 bales of cotton or it is placed in trailers and taken directly to the gin. A bale weighs around 1,200 pounds before the seeds are removed and around 500 pounds after the seeds are removed.

**FIGURE 10-17**
Cotton blossoms are first yellow, then turn pink. *Courtesy of James Strawser, The University of Georgia.*

**FIGURE 10-18**
Cotton is picked by large machines. *Courtesy of James Strawser, The University of Georgia.*

At the gin the seeds are removed and the cotton is cleaned of trash and foreign material. Once the seeds are separated and the cotton is cleaned, the lint is pressed into 500-pound bales, graded, and stored or sent to the mill. The seeds are a secondary industry. They are sent to a processor where they are pressed to remove the oil. The oil is used as a cooking oil or in a variety of products from margarine to salad dressing. The cake that is left over after the oil is pressed out is ground into a meal that is a valuable source of protein. Cottonseed meal is fed to cattle and other ruminants. It cannot be utilized by pigs and certain other simple-stomach animals because of a toxic substance in the seeds called gossypol.

## Flax

The use of linen dates back to the Stone Age when humans still wore skins for clothing but used linen for making fish nets. Over the years linen has been used for clothing in most countries of the world. Much of its popularity as a fabric for clothes was lost with the widespread availability of cotton that began in the early 1800s. Today it is still popular as a cloth for making tablecloths, napkins, and in recent years has regained some popularity as a clothing material.

Linen comes from fibers produced in a plant called flax that grows in climates with plenty of rain and no high temperatures. Most of today's flax is produced in Europe and New Zealand. However, substantial amounts are produced in North and South Dakota and Minnesota.

Plant stems contain two structures, the xylem and the phloem, that transport water and nutrients through the plant. Linen comes from the fibers (called bast fibers) that make up the phloem of the plant.

**FIGURE 10-19**
A high-quality cloth called linen is made from the flax plant. *Courtesy of the Flax Council of Canada.*

The plants grow to a height of about 3 feet and are harvested. The outer layer of woody material must be dissolved before the fibers can be removed. This is done by soaking the stems in warm water where bacterial action decays the material or in a more modern method where the covering is dissolved using chemicals. This process is called retting.

After the outer layer is removed the phloem fibers are passed through fluted rollers that break up the woody substances connected to the fibers. In a process called scrutching, the usable fibers are separated out. Then the fibers are combed out in much the same process used in carding cotton. This lays out the fibers parallel to each other so they can be spun into yarn.

The seeds of the flax plant are pressed for their oil. This valuable oil, called linseed, is used in the manufacture of paints and varnishes. As with the other oil seeds, the cake left from oil removal is used for livestock feed.

## Summary

The production of row crops in the United States is a gigantic business. No other country in the world can compare to the United States in terms of the amount or variety of crops that are produced. Much of the wealth of the nation is a result of our crop production. In the future perhaps new crops and new uses for old crops will be developed.

# 10 CHAPTER REVIEW

## Student Learning Activities

1. Examine the labels on foods in your home. List all the foods that contain grain or grain products. Remember that gluten and starch come from grain. Compare your list with others in the class.

2. Visit a grocery store and list all of the products that are made from wheat. Try to determine which type of wheat would be best suited for each product.

3. List all of the products in your home that were made from oil from seeds. Try to think of other ways oil from plants might be used someday. Share your ideas with the class.

## True/False

1. ___ Probably the very first crop that people planted was grain.

2. ___ Most of the wheat grown is used for human food.

3. ___ Most wheat is produced in the Northeast.

4. ___ More corn is produced in the United States than any other crop.

5. ___ The majority of corn production is in the Midwest.

6. ___ Much of the cloth produced today is manufactured from petroleum.

7. ___ Human-made fibers are more comfortable than fibers produced by agriculture.

8. ___ Cotton was probably first used in the Nile Valley in ancient Egypt.

9. ___ One of the most valuable products from peanuts is peanut oil.

10. ___ China produces over half of the world's supply of soybeans.

# Multiple Choice

1. A large portion of our agriculture industry is
   a. row cropping.
   b. corn production.
   c. fruit production.

2. The grasses that produce the most and best seeds are selected for
   a. grazing.
   b. cultivation.
   c. flax.

3. The oldest of the major grain groups is perhaps
   a. wheat.
   b. rice.
   c. corn.

4. Wheat flour is relatively high in
   a. fat.
   b. sugar.
   c. protein.

5. Gasohol is a mixture of gasoline and
   a. ethanol.
   b. diesel fuel.
   c. propane.

6. The two most widely produced oil crops are soybeans and
   a. corn.
   b. olives.
   c. peanuts.

7. A shrub-like plant that requires a long growing season and warm temperatures is
   a. flax.
   b. cotton.
   c. linen.

8. Linen comes from fibers produced in a plant called
    a. gluten.
    b. cotton.
    c. flax.

9. Plant stems contain these two structures
    a. xylem and phloem.
    b. xylem and gluten.
    c. phloem and gluten.

10. The seed of the flax plant is pressed for
    a. gluten.
    b. linen.
    c. oil.

# Discussion

1. What factors determine what row crop is grown in each state?

2. What are the three classifications of crops?

3. What are the types of wheat?

4. Why is hard red winter wheat so popular?

5. Durum wheat is used for what purpose?

6. Other than color, what is the difference between white and yellow corn?

7. What are some uses for corn?

8. What functions are served by the xylem and phloem?

9. Where is the majority of corn grown? Why?

10. What uses are made of soybean oil?

# Forest Science

## Student Objectives

*When you have finished studying this chapter, you should be able to:*

- Discuss the extent of the forest industry.
- Explain why forestry can be considered agriculture.
- Explain the concept of natural succession.
- Discuss the forest as an ecosystem.
- Discuss the importance of the forests to humans.
- Discuss the production steps in growing trees.

One of the largest industries in agriculture is **forestry.** At one time forestry was considered a natural resource. Giant forests once stretched across the country and people made use of the trees for building homes. When it became apparent that these forests would not last forever, the cultivating of trees began. Trees are now grown very much the same as crops are grown. They are planted, managed, and harvested. Once harvest is complete new trees are planted and a new crop is grown.

FIGURE 11-1

Like any crop, trees are planted, managed, and harvested. *Courtesy of James Strawser, The University of Georgia.*

The forestry industry stretches from coast to coast and encompasses almost 500 million acres of trees. From these trees come thousands of different types of products that are manufactured using wood fiber. Over a million and a half people are employed in the management, harvesting, processing, and marketing of these products.

Some people think that the harvesting of timber is a horrible practice that destroys nature. This is an erroneous concept. Because of the conservation efforts of the forestry industry and the U.S. Forestry Service, today there are more trees in this country than there were 100 years ago. Each year close to two billion trees are planted in the United States. This represents more than six new trees a year for every American. Through the replanting of the forests, new plant growth flourishes. Each year about 27 percent more trees are planted than are harvested. Areas that were once used for row cropping are now used for growing trees. Hilly land not suited for farming with modern machinery is now planted in pine trees and is once again productive land.

Wildlife greatly benefits from the forest industry. As new trees are planted, lush growths of plants

FIGURE 11-2

Each year about 27 percent more trees are planted than are harvested. *Courtesy of Georgia Pacific.*

grow with the seedling trees. These plants provide food and habitat for wildlife. Several species of animals have been reintroduced where they had ceased to exist. For example, the white-tailed deer has made a tremendous comeback in recent years. Some sources say that there are now more deer in this country than before the first European settlers arrived. A few decades ago, the wild turkey was on the verge of extinction due to the loss of its habitat and overhunting. Today there are over four million wild turkeys living and thriving in our forests.

Most of the trees that are harvested in the United States were planted or have grown up from forests that were cut. Only in the Pacific Northwest are there large tracts of old-growth timber. This timber is also call virgin timber and has never been cut. Even today, timber companies are cutting old-growth timber in the Pacific Northwest. No finer construction timber exists. Homes all over the world are made from lumber cut from the trees in the Pacific Northwest. Many people object to cutting the huge old trees. A constant struggle seems to be ongoing between preservationists and the logging industry. The preservationists insist that as much of the old-growth timber as possible must be preserved. The logging industry points out that there are over four million acres of old-growth timber preserved in this country that, by law, can never be cut. They contend that the country's housing industry needs the material from the large trees that are in the areas where harvesting is permitted. They further argue that if logging is stopped many jobs will be lost as a result.

Another point of controversy is over **wildlife habitat.** A good example is the spotted owl that lives in the Pacific Northwest. Some biologists contended that the owl needed large tracts of old growth timber in which to breed and raise their young. Logging operations were halted and many

FIGURE 11-3
Today there are more of some types of wildlife in the forests than ever before. *Courtesy of Vann Cleveland, Progressive Farmer.*

FIGURE 11-4
Some object to the cutting of large, old-growth trees. *Courtesy of Jim Rathert, Missouri Conservation Department.*

FIGURE 11-5
Logging operations were stopped in efforts to save the spotted owl.
*Courtesy of Northwest Forestry Association and Evergreen Foundation.*

people lost their jobs in the effort to preserve the owl. The issue was debated over whether or not preserving the owl was more important than providing jobs. Later evidence showed that the owls could reproduce in second growth forests.

## The Natural Forest

At one time, trees cut for lumber came from the natural forest. Different types of forest grew in different regions of the country. Basically, there are two classifications of timber: **hardwood** and **softwood.** A hardwood is a tree that is a broad leaf deciduous (sheds its leaves in the winter) tree. A softwood is a cone-bearing, evergreen tree called a conifer. Actually, the classification has little to do with the hardness or softness of the woods. Some hardwoods, like yellow poplar, are softer than softwoods like southern pine.

The upper regions of the Northeast at one time were covered in vast hardwood forests. Throughout the Midwest and the upper regions of the South, hardwoods also grew. Along the coastal plains of the South, the predominate trees were pines. The Rocky Mountain region was covered with conifers such as lodge pole and ponderosa pine. On the West Coast, gigantic softwoods such as Douglas fir, spruce, and redwood trees covered the slopes.

## Forest Succession

The reason certain types or species of trees dominate a region is because of a natural process called **succession.** When a forest first begins from bare ground as a result of fire, clear-cutting, or other causes, a lot of different plants germinate. These plants all compete for water, nutrients, and sunlight. The trees grow taller than other plants so

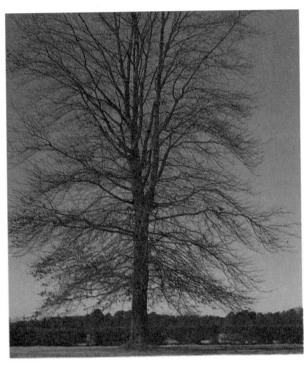

**FIGURE 11-6**
A hardwood is a deciduous tree. *Courtesy of James Strawser, The University of Georgia.*

**FIGURE 11-7**
A softwood is an evergreen, cone-bearing tree. *Courtesy of James Strawser, The University of Georgia.*

eventually trees shade the other plants. Plants that do not receive enough sunlight generally stop growing and die. Trees that grow well in a particular climate are usually the ones that grow faster than the other species of trees. Also, some species just naturally grow faster than other species. The limbs of the faster growing trees eventually form what is called a **canopy** or overstory above the other trees. This prevents trees of the same species from growing in competition. Trees that are not tolerant of shade will not grow under the canopy. However, trees that are tolerant to shade continue to grow at a slow pace.

Eventually the trees that are shade tolerant may overtake and suppress the trees that first created the overstory. This is particularly true if the first trees mature and die. A good example of this pro-

**FIGURE 11-8**
The limbs of faster growing trees unite to form a canopy. *Courtesy of Jim Rathert, Missouri Conservation Department.*

## Career Development Events

The Forestry Field Day Career Development Event is a team competition. Team members train for skills used in today's forestry industry. Tree planting, forest management, tree disorders, timber stand improvement, and estimating pulpwood and saw timber volumes are a few of the skills. Many forestry management practices are included in the FFA Forestry event. Contestants must understand and answer forestry management questions concerning stand classification, merchantable trees, basal area, and cutting practices. Team members must practice and be ready to give estimations of pulpwood and saw timber standing in a given area by measuring the tree heights and diameters. Testing of the contestant's ability to improve a timber stand by harvesting select trees due to insect infestation, disease, lack of vigor, or tree conformation is another part of the event.

Foresters must develop good math skills. Many measurements and calculations are required to manage timber land efficiently and effectively. The event tests the student's ability to make ocular estimates of trees, calculate the number of acres and placement of boundary lines in a given area. The FFA team must read a compass and decide points and distances during the compass course. Members must identify not only trees but the insect and diseases that attack the trees. The FFA Forestry Event requires a winning team to spend time

Forestry events allow students to display their skills in forestry measurement and management. *Courtesy of Blane Marable, Morgan Co. High School, Madison, GA.*

studying classroom materials and practicing skills. State winning teams compete at the National Forestry Contest.

The Forest Management Proficiency awards encourage the use of forestry practices that conserve and increase the productivity and economic value of land. The award is presented to individual FFA members who have excelled in experiences using the principles and practices of today's forestry industry. To apply the student must be involved in a forestry related Supervised Agricultural Experience Program (SAEP). The student must maintain a list of forest management skills learned and practiced during the project. The student must also develop an inventory of buildings, equipment, tools, and materials used in the experience program and keep accurate records of income and expenses of forestry products produced, harvested, and sold. By keeping a list of accomplishments such as land preparation, planting, seeding, thinning, and harvesting, the student can show how he or she has improved the land use for forestry products.

Maria won the forestry management proficiency award in her FFA chapter. Her program in forestry developed from her interest in her agriculture forestry class and part-time job at a tree nursery. Over a four-year period, Maria recorded 856 hours working for the nursery and developed skills in customer service, propagation, planting and maintenance of trees. Her experience at the nursery gave her the idea to propagate and grow Leyland cypress as Christmas trees. At her home she rented an acre of land from her parents and planted 200 1-gallon Leyland cypress trees that she bought from her employer. The first year she also built a 10′ x 15′ cold frame and began propagating cuttings. During the second and third years, she pruned, fertilized and kept weeds away from her trees and sold rooted cuttings to local nurseries. The fourth year, Maria had about 50 trees that were large enough to sell. She pruned and sprayed the trees, preparing them for the Christmas season. The first week of December she cut the trees and sold them at the local farmers' market for $25 each. The fifth year she sold 87 trees and replanted the trees with two gallon plants that she had propagated earlier. By her senior year in high school, Maria had an annual net income of over $6,000. Her agriculture class, part-time job, and supervised agriculture experience provided the knowledge, experience, and opportunity to profit from her interest of growing trees.

cess is the forests that covered the upper South. The first trees to dominate were the fast-growing pines such as the loblolly, slash, and shortleaf. These species created canopies or overstories that suppressed the species of trees that were not tolerant of shade. However, beneath this canopy, hardwood species such as red oak and white oak grew. These species tolerate shade fairly well and will grow beneath the pine canopy. A characteristic of the southern pine is that it will mature and die at an earlier age than the oaks. When a pine dies, the space in the canopy may be filled with a hardwood such as oak. When the hardwoods create a canopy, little sunlight will penetrate. Since pines are intolerant of shade, they do not grow well under the oaks. The hardwoods eventually become the dominant species. This is called the climax vegetation, because it is unlikely that other species will overtake the hardwoods. Succession is a very slow process that may take 200 years to reach the climax vegetation stage. When the Europeans explored this region of the continent they found a mixture of

**FIGURE 11-9**
Hardwoods eventually become the climax vegetation. *Courtesy of James Strawser, The University of Georgia.*

pine and hardwoods. The farther south they went, the more pine they found because the milder climate and sandy soil of the coastal plains favored the pines.

## The Forest Ecosystem

Forests play an important part in our lives because they are a part of our ecosystem. An ecosystem is an area where all the living and nonliving features of the area depend on each other. The forests of the world are instrumental in the support of life on the planet. Plants use carbon dioxide and give off oxygen in their life processes. Since animals use oxygen and give off carbon dioxide, plants support all animal life, including humans. Forests not only provide oxygen, but also aid in removing air pollution.

Young rapidly growing trees produce more oxygen than older trees. As a tree produces new wood, carbon dioxide is removed from the air and oxygen is added. According to The American Forest and Paper Association, in the period of one year, an acre of healthy trees may produce 4,000 pounds of wood, use 5,889 pounds of carbon dioxide, and give off 4,280 pounds of oxygen. Since we have almost 500 million acres of trees, this represents a lot of oxygen that is put in the atmosphere.

Forests provide us with clean water. Roots of trees prevent water from running rapidly down a slope and instead cause the water to soak into the ground, where it accumulates. This recharges streams, lakes, and wells without adding the soil from runoff and helps keep the soil in place on slopes. In doing this, trees prevent lakes and streams from being muddy. Muddy lakes and streams prevent fish and other aquatic life from growing and lower our supply of clean drinking water.

Forests provide habitat for animals and other plants. Many species of plant life live beneath the

**FIGURE 11-10**
Young rapidly growing trees produce more oxygen than old trees. *Courtesy of Georgia Pacific.*

**FIGURE 11-11**
Forest vegetation helps keep streams clean. *Courtesy of James Strawser, The University of Georgia.*

**FIGURE 11-12**
Loggers have to abide by regulations regarding timber harvest. *Courtesy of James Strawser, The University of Georgia.*

canopy of trees where they are sheltered from direct sunlight. These plants would not thrive in the open sun. Many species of wildlife inhabit the forests and depend on the trees for their existence. Different types of forests support different types of wildlife. Animals and birds prefer particular types of foliage for food or to make nests in. Without this habitat certain species would become extinct in a short while. The federal government has established laws and regulations regarding the protection of animal and plant species in the forests. Loggers and timber companies have to abide by the regulations in the production and harvesting of timber.

The United States is one of the most beautiful countries in the world. In large part this is due to our forests. Millions of acres of land are set aside by state and federal government to protect the natural beauty of our forests. These national and state forests preserve a portion of the virgin forests that were here when the first settlers arrived. Today these areas provide a valuable resource in terms of recreation. Hiking, fishing, hunting, and camping

**FIGURE 11-13**
Our forests provide us with areas for recreation. *Courtesy of Vick Christ.*

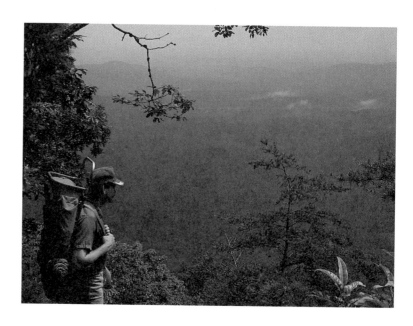

are favorite pastimes of many Americans and our forests make these hobbies possible.

One of the greatest contributions our forests make is the large amount of jobs associated with the forestry industry. Billions of board feet of lumber are grown, harvested, and processed each year. This requires a gigantic workforce to care for the trees, remove them from the forest, and convert the logs into finished products. This helps keep the nation's economy sound. It also supplies us with essential products that are used in almost all aspects of our lives.

## Wood Fiber Production

One of the largest segments of agriculture is the planting, management, and harvesting of trees. Americans at one time thought that the supply of timber was inexhaustible. Then around the turn of the 20th century, people began to realize that it was possible to run out of forests to harvest if measures were not taken.

**FIGURE 11-14**
Around the turn of the century people realized that the forest would one day be depleted unless actions were taken. *Courtesy of Northwest Forestry Association and Evergreen Foundation.*

Many people also thought that the methods used to harvest timber were wasteful and led to the depletion of the soil. As a result, in 1891 Congress passed the Federal Forest Reserve Act. This legislation gave the President the power to set aside timber lands to be managed so that people in the future would have forests to use and enjoy. Later, large tracts of land were designated as federal timber lands. The National Forest Service was established to manage these lands. Legislation over the last one hundred years has set aside land as being publicly owned. Some land is designated as national parks and is used only for recreational purposes. Other land was set aside as **wilderness areas** where the environment is carefully preserved so that it will remain as close to its original natural state as possible. Still other public timberland is harvested by companies who buy the rights to cut the timber.

Forestry was once looked on as being a natural resource that could be utilized. However, in recent years the timber industry has become an agricultural enterprise. Trees are planted, cultivated, and harvested in much the same manner as other agricultural crops. This means that instead of being a resource that will one day be exhausted, the timber industry can be sustained year after year through the careful planting and cultivation of trees.

For many years, areas clear-cut of all or most of the trees have been seeded or set out with tree seedlings. This allows the forest to grow back in much the same manner that it originally did. Throughout the Northwest, forests have long been regenerated by scattering seeds on newly cut forest areas or by setting out small trees. For a forest to grow from seedlings to harvestable timber may take from 70 to 100 years. For this reason much of the timber industry has moved to the Southeast. In fact, the leading producer of wood is not Washington or

**FIGURE 11-15**
A large portion of the trees harvested in the South go into plywood. *Courtesy of American Plywood Institute.*

Oregon, but Georgia. In this region, the hot humid weather allows the trees to grow in much less time. A carefully managed stand of timber can go from seedlings to harvestable trees in as little as 15 years. It must be remembered, however, that the timber from the Northwest provides a higher quality lumber for the construction industry. A large portion of the trees harvested in the South go for making paper or for plywood core material.

## Tree Farms

**Tree farms** are operations where trees are planted in rows and are given care throughout their growth period. At the proper time they are all harvested. New trees are then planted and the crop starts over again. There are tree farms in almost all states. The most are located in the South.

The process of tree farming begins in the nursery, where superior seedlings are produced for tree plantations. Research goes into the selection and production of seeds from high-quality trees that are resistant to diseases and grow rapidly. Many of the seedlings are crosses between two species or

**FIGURE 11-16**
Superior tree seedlings are produced in nurseries. *Courtesy of Georgia Pacific.*

strains of trees. For example, a longleaf pine may be crossed with a loblolly pine to produce a tree that is superior to either of the parent trees. This cross, known as a hybrid, benefits from a phenomenon known as heterosis or hybrid vigor. The hybrid grows more rapidly and is generally healthier than the purebred strains of trees.

After the trees are set out in large areas called plantations, they are managed using techniques that interrupt the process of succession. The underbrush may be cleared by using mechanical means, chemicals, or by prescribed burning. The use of fire in prescribed burning is aimed at destroying vegetation that competes with the trees for water and nutrients. The fire is carefully controlled to prevent damage to the trees and to prevent the fire from escaping.

Succession is also controlled by thinning the trees. As the trees grow and begin to compete with each other for sunlight and nutrients, some of the trees are harvested. Usually the trees that are cut are those that are slowest growing or poorest in quality. The best trees are kept to grow. The trees harvested in the thinning process are used for fence posts, poles, or paper. As the trees grow and mature they are harvested and used for making paper, lumber, and other uses.

## Summary

Forestry is a large part of the agricultural industry. Trees are grown and harvested as a crop. If we continue to replenish the forests we will always have wood to use to make the many products we enjoy. At the same time, we can enjoy the beauty of green forests.

# 11 CHAPTER REVIEW

## Student Learning Activities

1. Take the side of a preservationist or a logger in a controversy over the proposed cutting of a tract of timber. Prepare your points and arguments and debate someone with the opposite view.

2. Make a list of all of the products in your classroom that are made from trees. Do not forget the paper products. Compare your list with others in the class.

3. Select a particular type of tree that is grown for timber. Research the types of products that are made from the trees. What are the characteristics and uses of the lumber from the trees? Report to the class.

# True/False

1. ___ One of the largest industries in agriculture is forestry.

2. ___ There are more trees in the United States today than there were 100 years ago.

3. ___ Wildlife greatly benefits from the forest industry.

4. ___ Plants use oxygen and give off carbon dioxide in their life processes.

5. ___ Forests aid in removing air pollution.

6. ___ Most tree farms are located in the North.

7. ___ Usually the trees that are cut are those that are growing best.

8. ___ Succession is controlled by thinning the trees.

9. ___ Preservation activists insist that as much of the old growth timber as possible must be preserved.

10. ___ Succession is a very fast process.

# Multiple Choice

1. At one time forestry was considered a
   a. conservation effort.
   b. a bad investment.
   c. natural resource.

2. The forestry industry stretches from coast to coast and encompasses around
   a. 500 million acres of trees.
   b. 100 million acres of trees.
   c. 5 million acres of trees.

3. Each year there are close to _____ billion trees planted in the United States.
   a. one
   b. two
   c. five

4. The reason certain types of trees dominate a region is because of a natural process called
    a. deciduous.
    b. succession.
    c. overstory.

5. An area where all living and nonliving features depend on each other is called an
    a. ecosystem.
    b. oxygen system.
    c. industry.

6. The legislative act giving the President power to set aside timber lands is called the
    a. American Forest Act.
    b. U.S. Forestry Service Act.
    c. Federal Forest Reserve Act.

7. Young rapidly growing trees produce more _____ than older trees.
    a. carbon dioxide
    b. oxygen
    c. pollution

8. One of the greatest contributions our forests make is the large amount of _____ associated with the forestry industry.
    a. jobs
    b. carbon dioxide
    c. hunting

9. The leading producer of wood is
    a. Washington.
    b. Oregon.
    c. Georgia.

10. Operations where trees are planted in rows and are given care throughout their growth period are called
    a. nurseries.
    b. loblollies.
    c. tree farms.

# Discussion

1. How does wildlife benefit from the forest industry?
2. What is the difference between hardwood and softwood timber?
3. What is a canopy?
4. What are the benefits of forests for animals?
5. What is the Federal Forest Reserve Act?
6. Which state is the leading producer of wood? Why?
7. Why is the harvesting of timber not really a threat to nature?
8. What is virgin timber?
9. What are the benefits of forests for humans?
10. Why did the cultivation of trees begin?

# Protecting the Environment

## Student Objectives

**When you have finished studying this chapter, you should be able to:**

- Explain why everyone should be concerned with the environment.

- Discuss the importance of agriculture to a growing population.

- Tell how pesticides have the potential to cause harm to the environment.

- Explain how the threat of harm from pesticides is minimized.

- Discuss the difference between point and nonpoint pollution.

- Tell how agriculturists work to prevent water pollution.

- Explain the importance of wetlands.

Everyone should be concerned with protecting the **environment.** There is only a limited amount of space on our planet. Humans continue to multiply and the population grows. This means that there is less space for each person. At one time in history the population was so small that when conditions became bad, people could move to a new place. The places where people can move to a fresh environment are almost gone. Scientists predict that by the year 2050 there will be over 11 billion people in the world. That is over twice the current population. Most of you will be alive at that time and will have to cope with the large population.

Since there will be less space for each of us, it becomes increasingly important that we care for the environment. The quality of life will be directly related to the conditions of the environment. Clean water to drink and clean air to breathe are essential. Most of the beauty we enjoy comes from the nature surrounding us. Few things are more pleasing than a green forest or a clear stream. The beauty of the surroundings will have to be protected from destruction. The more people living in the environment and the more people using our resources the greater the likelihood that the environment will become polluted. Resources are those things we have available for our use. **Natural resources** occur in nature and they include water, air, mineral deposits, and soil. Usually when these resources are used up, they are gone forever. **Renewable resources** are resources

**FIGURE 12-1**
Most of the beauty we enjoy comes from the nature surrounding us. *Courtesy of Ray Herren, The University of Georgia.*

that can be replaced, such as crops. These crops may include trees, flowers, and plants grown for food. In the future we will have to rely more heavily on renewable resources supplied by agriculture.

Not only will there be less space for all of us, but all of the people will need food to eat. Agricultural systems all over the world will have to supply the needs of over 11 billion people—and that number will continue to rise. With all these people needing space in which to live and work, less land will be available to grow food. Agriculturists will have to produce much more food on less land. Because of this, farming practices will have to become more intense, which means getting more production out of fewer acres of land. This will be a challenge to produce enough food and fiber to feed and cloth the world's population without wearing out the land or polluting the environment.

Agriculture always has and always will have an effect on the environment. In order to grow crops, forests have to be cleared, and less room is available for wildlife. To produce crops efficiently, fertilizers and pesticides have to be applied. The constant tilling of the soil can cause erosion. Animals that are raised in confinement produce a lot of manure that has to go somewhere. The answer cannot be the elimination of agriculture. We must have food, shelter, and clothing. The answer is that we have to design our agricultural production methods so that they will protect the environment.

## Pesticides

Pesticides are substances that kill pests. Each year about 570 million pounds of pesticides are used to aid in the control of insects, weeds, and fungi. This may seem like a lot of chemicals but you must realize that there are over 300 million acres of crop land in the United States. People have developed negative

FIGURE 12-2
With a huge and growing population, agriculturists will have to produce more food on less land. *Courtesy of James Strawser, The University of Georgia.*

FIGURE 12-3
The constant tillage of the soil can cause erosion. *Courtesy of James Strawser, The University of Georgia.*

attitudes over the use of pesticides. Some view their use as being terribly destructive to the environment.

In the past there have been problems with pesticides. Years ago a very popular insecticide was a chemical called DDT. This pesticide was very effective in killing insects and solved a lot of problems all over the world. However, DDT was very residual. This means that it stayed in the soil for a long time before it broke down. This caused problems with the food chain. Birds that ate insects affected by the insecticide retained DDT in their bodies. Water birds might eat fish that had consumed insects or other aquatic life affected by DDT. The DDT built up each time the birds ingested the chemical. Problems developed in the shells of the eggs the birds laid. The shells were so thin that the eggs would not hatch. This interrupted the life cycle of the birds. DDT also turned up in other forms of life. The chemical was passed up the food chain and all of the animals in the chain were affected. Because of these problems, DDT and chemicals like it have been banned. Modern pesticides break down very rapidly and pose much less danger than the older pesticides.

Sometimes, problems with agricultural chemicals are blown out of proportion. For example, in

**FIGURE 12-4**
The chemical Alar was reported to be a carcinogen to laboratory animals. *Courtesy of Dow/Elanco-French Studios.*

1989 the media began a series of stories on the chemical daminozide, known by its trade name, Alar. This chemical acts as a growth regulator that is used on apples to prevent the apples from falling off the tree too early. This helps the apples to grow into a better shape. Also it prolongs the storage life. According to the reports, the chemical proved to be a **carcinogen** (a substance that causes cancer) when fed to laboratory rats. As a result, near panic ensued among consumers. Apples and apple products were taken off the market. Tons of apples were dumped as garbage and the fruit was banned from school cafeterias. The apple industry suffered tremendous losses. The money earned from the sale of apples fell by more than $100 million. In addition, millions of dollars of income was lost from all the support services of the apple industry. Many apple producers reached the brink of bankruptcy. This example points to the concern of the public about the safety of food.

Further investigation showed little justification for the panic. In the study, laboratory rats had been given enormously high amounts of Alar before they developed tumors. At much lower rates the chemical had no effect at all. According to the American Council on Science and Health, a person would have to eat the equivalent of 28,000 pounds of apples per day for 10 years to equal the amount of Alar fed to laboratory animals before they developed tumors. The amount of Alar that was on the apples was almost unmeasurable. In no way could a human eat enough apples in a lifetime to consume enough Alar to cause problems. Yet in spite of this fact, the chemical was pulled off the market.

The use of pesticides are necessary if we are to provide the population with high-quality food. Without them both the quantity and the quality of food would be lower. Produce that is damaged by insects or disease will be of such low quality that

FIGURE 12-5
A person could not eat enough apples in a lifetime to receive the amount of Alar given to the lab animals. *Courtesy of James Strawser, The University of Georgia.*

**FIGURE 12-6**
Pesticides are necessary to provide wholesome food. This fruit is not edible because of pest damage. *Courtesy of James Strawser, The University of Georgia.*

consumers will be reluctant to accept it. Disease causes bad spots on fruits and vegetables. Fruits and vegetables that have been partially eaten by insects will not only look bad and taste bad but will also spoil more easily.

Modern pesticides are safe to use. The United States Department of Agriculture (USDA) and the Environmental Protection Agency **(EPA)** have strict regulations regarding the development and use of pesticides. Before a pesticide can be released for use, it must undergo years of rigorous testing to ensure it is safe. It must not only pose no threat to humans but must not be harmful to the environment. Modern pesticides are not only necessary but safe if used properly.

## Water Pollution

One of the most precious natural resources we have is our water. This compound is essential for life. Without it humans would die within a few days. We depend on it to sustain the lives of the plants and animals we grow for our food supply. Most of our industries require large amounts of water, and so does much of our recreation. Fishing, boating, and

FIGURE 12-7
We depend on clean water to sustain life. *Courtesy of James Strawser, The University of Georgia.*

swimming require a huge amount of clean water. We have only a certain amount of water. Water is used and recycled over and over again through a process known as the **hydrological cycle.** Water evaporates, condenses and returns to the earth as rainfall.

Rainfall replenishes two types of water storage—surface water and groundwater. Surface water is the water that is in streams, lakes, and the ocean. When rain falls, the amount that is not absorbed into the ground flows downhill into streams. Small streams flow into larger streams and the large streams eventually flow into the oceans. A few years ago surface waters in this country faced serious pollution problems. Many streams and lakes were so polluted that very little life could exist there.

The other type of water is called groundwater. Groundwater is the water that is stored under the surface of the ground. When rain hits the ground, some runs off and some is absorbed into the ground and accumulates in storage areas called **aquifers.** Aquifers are formations in sand, rock, or gravel that hold water. Wells are drilled into aquifers to retrieve water. A large portion of our drinking water comes from aquifers.

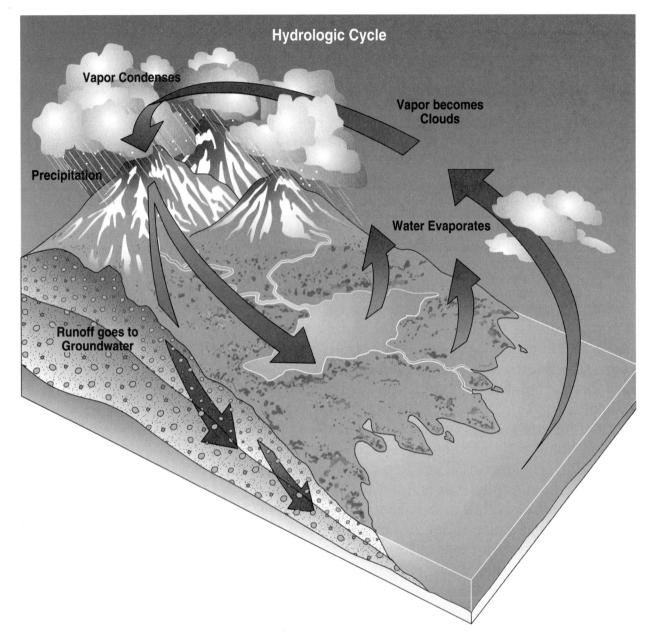

FIGURE 12-8 Through the hydrological cycle water is constantly recycled.

Groundwater pollution can be serious. Fortunately, as water travels through the ground, it is filtered and most pollutants do not reach the groundwater. However, where accidental chemical spills occur, problems can happen. However, the EPA has determined that fewer than 1 percent of

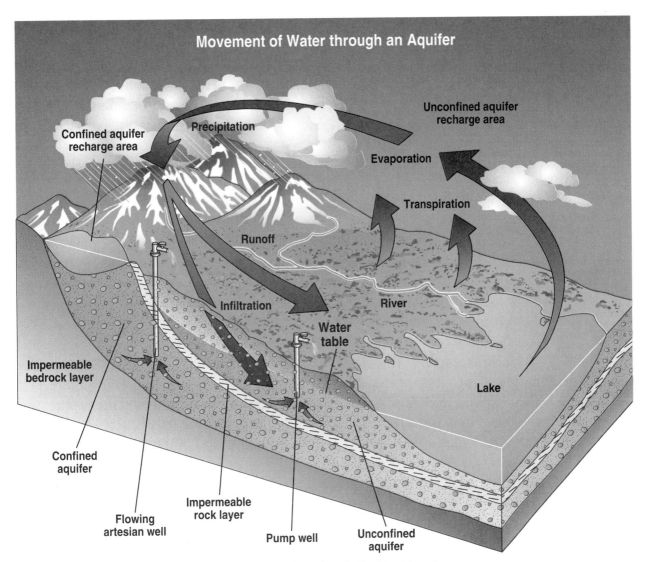

**Movement of Water through an Aquifer**

Confined aquifer recharge area

Precipitation

Unconfined aquifer recharge area

Evaporation

Transpiration

Runoff

Infiltration

River

Water table

Lake

Impermeable bedrock layer

Confined aquifer

Flowing artesian well

Impermeable rock layer

Pump well

Unconfined aquifer

FIGURE 12-9  Aquifers are formations in sand and rock that hold water.

the rural domestic and community water systems contain unacceptable levels of pesticides.

Areas that have shallow aquifers or coarse, sandy soil are in more danger of pollution. The deeper the water supply and the heavier the clay content of the soil, the safer the water is. This is because the depth and the clay slows down the movement of the water. This gives more time for the water to be filtered. New regulations governing the use of pesticides have lessened the likelihood of pollution of the groundwater.

In 1970, President Nixon signed into law the National Environmental Policy Act, which led to the establishment of the Environmental Protection Agency (EPA). This agency was charged by

## Career Development Events

**FFA**

In every community across the nation there is a need for improvement. The FFA program encourages students to identify community needs and take action. Through the years community programs have developed active, experienced, and knowledgeable leaders by educating and involving students. Cities, towns, and communities across American have benefited from these projects. Projects have created and expanded community services and improved the quality of life. Community development is one of the major areas addressed in a local FFA chapter program of activities. To make a meaningful impact upon the community the whole chapter should be involved in the planning and action. The first step is to select a chairperson and committee. The chairperson will lead the chapter in the planning, initiation, and public relations for the effort. The committee will identify needs through questionnaires,

attending public meetings, or surveys of the community. After collecting the information, the committee will assess the information and identify a community project. Identification of resource persons and community groups for support of the activity is important. The involvement of groups such as FFA Alumni, Chamber of Commerce, and civic clubs can provide not only financial support, materials, and equipment but offer expertise in needed areas. Cooperative efforts with other community groups will increase the success of the project. A plan of action is necessary to organize the phases of the project and its completion. A step-by-step project plan, work schedule, budget and available resources should be included in the plan of action. After developing the plan of action it is time to start the project. Involving all FFA members and interested community groups will result in a successfully completed project. Publication of community development activities through newspapers, radio,

Congress to "protect the nation's land, air, and water systems." Policies, laws, and regulations have been and continue to be established to prevent pollution and to clean up pollution that has already occurred. Since 1970 a lot of progress has been made in cleaning up the air and water. Many bodies

and television will increase community awareness and support. After the project is completed a community-wide meeting, such as an FFA banquet, should recognize those groups and individuals who participated. You can make a real contribution to your community!

In competing in the Building Our American Communities (BOAC) Award, students help to improve the environment. *Courtesy of Blane Marable, Morgan Co. High School, Madison, GA.*

of water that were once considered dead now have been brought back to life by cleaning up the water and stopping pollution.

Pollution comes about as a result of contaminants entering the water in two main ways: point and nonpoint sources. A **point source** means pollution from a specific place. For instance, a factory that dumps waste in a river is a point source of pollution because the contaminants can be traced to a single point. The EPA has made tremendous strides in halting point source pollution. Industries have been required by law to stop dumping or allowing contaminants to enter waterways. Water used in industry and returned to streams or lakes is carefully monitored to ensure that it contains no polluting substances.

Pollution from **nonpoint sources** is more difficult to deal with. Nonpoint means that the pollution does not come from a single point but from a wide area that is difficult to identify. The pollution is usually from a number of sources that collectively affect the environment. Agriculture is the largest origin of nonpoint pollution in the United States. This is due in part to the enormous size of the agricultural industry, where millions of acres of land are in production. Each year this land receives

**FIGURE 12-10**
The EPA requires that water returned to a stream be clean.
*Courtesy of James Strawser, The University of Georgia.*

treatment in the form of plowing, cultivating, storage of animal wastes, and the application of pesticides. As a result, agricultural pollution comes from several sources: fertilizers, pesticides, animal wastes, and soil erosion. Recently, laws have been passed to help protect the environment from these pollutants.

## Soil Erosion

Our largest environmental problem stemming from agriculture is soil erosion. Erosion is the wearing away of the soil by wind or water. This problem is two-fold; the loss of top soil and the pollution of surface water by the displaced soil.

Research indicates that the sediments from soil cause around $6 billion in damage each year. Almost all improved land in the United States has a system of drainage ditches that carries off excess water. Eroded soil clogs these waterways and has to be periodically removed. Similarly, navigable waterways can become so clogged that they have to be dredged out in order for the vessels to pass through. This is a tremendously expensive process that takes a lot of time and resources.

FIGURE 12-11
Soil erosion is our most serious pollution problem stemming from agriculture. *Courtesy of USDA Natural Resources Conservation Service.*

**FIGURE 12-12**
Some species of aquatic life cannot tolerate muddy water and die from the lack of oxygen. *Courtesy of USDA Natural Resources Conservation Service.*

Another serious problem caused by sediment pollution is the loss of fish and other aquatic life in lakes and streams. Some species cannot tolerate muddy water and die from lack of oxygen or lack of clean water.

As the agricultural industry has developed, methods of protecting the soil from erosion have developed also. Intensive efforts began back in the 1930s with the creation of the Soil Conservation Service (SCS). The mission of this agency was to help prevent the loss of soil and to stop pollution caused by erosion. The SCS has made tremendous strides in the control of erosion. The name was changed in the early 1990s to the Natural Resources Conservation Service (NRCS). Modern efforts are aimed at keeping as much vegetation on the ground as possible. Plants hold the soil together by creating a support system for the soil particles. Roots penetrate the soil and through their fibrous structure hold the soil in place. Many crops are now grown with a bare minimum of plowing. This not only allows more plant roots and plant cover to hold the soil but also slows the flow of water over the land. In some areas, crop residues were plowed under but

**FIGURE 12-13**
In some areas the soil is tilled as little as possible to prevent soil erosion. *Courtesy of James Strawser, The University of Georgia.*

now are left until spring. The new crop is planted in the stubble left from the previous crop.

New varieties of grasses and other plant covers are also being developed. Bare hills are planted in grasses that grow quickly, have an expansive root system, and do well on eroded soils.

## Nitrate Pollution

Water is also polluted by agricultural production through the use of commercial fertilizers and animal wastes. Animals that are raised in confinement produce a lot of manure in a concentrated area. This material is generally used on fields as a fertilizer. Within both fertilizers and manures are plant nutrients called nitrates. In surface water, a major problem is that nitrates cause plant life in streams and lakes to flourish. This may not seem like a serious problem, but too much plant matter in the water can cause trouble. Clogged waterways are difficult for boats to maneuver through and interferes with recreational activities. Also, when the plant material dies and decays, oxygen is taken from the water. When a large amount of vegetation decays,

**FIGURE 12-14**
Animals in a confinement operation produce a lot of manure. Most of it is spread on fields as fertilizer. *Courtesy of James Strawser, The University of Georgia.*

FIGURE 12-15

A lagoon is a pond of water into which waste material like manure is directed. The material decomposes in the water. *Courtesy of James Strawser, The University of Georgia.*

much of the oxygen can be removed from the water and the fish may die.

Two practices can help control the problem. The first is to control the runoff from rainwater. Diversion ditches are cut into the area surrounding the livestock facility to prevent runoff containing waste materials from getting into streams. The drainage ditches divert the runoff into holding areas where the risk from runoff is at a minimum.

The second practice is to use a lagoon as a holding tank for the manure. Basically a lagoon is a pond of water into which the waste is directed. Here the material is decomposed by anaerobic bacteria. These bacteria live beneath the surface of the water in the lagoon and thrive without oxygen. Anaerobic bacteria break down the waste material and convert the nitrogen into a gaseous state. At the proper stage of decomposition, the material is pumped from the lagoon and spread on the fields.

## Wetlands

Many areas in our country are known as **wetlands.** A wetland is an area where the ground is naturally

**FIGURE 12-16**
Scientists now know that wetlands provide several essential roles in the ecosystem of an area. *Courtesy of James Strawser, The University of Georgia.*

saturated for much of the year. These areas were once thought to be wastelands that served no good purpose and that should be drained to be productive. It is estimated that over half of the wetlands of the United States have been destroyed for construction or for agricultural purposes.

Scientists now know that wetlands play several essential roles in the ecosystem of an area. These swampy areas help to filter water before it reaches open areas that store surface water. Earlier in this chapter, problems with chemicals, sediments, and other pollutants in runoff water were discussed. Wetlands catch runoff water before it reaches deeper bodies of surface water. The vegetation removes nutrients and other pollutants from the water and as a result the open areas of surface water are much cleaner. Wetlands also act as a sponge or storage area that collects and absorbs flood water before it can damage crops, homes, or other property.

Wetlands are critical to the life cycle of many different plants and animals. They may provide water for drinking, food, or shelter for a variety of animals in the area. Also, some plant species will only grow in areas where the ground remains saturated for

**FIGURE 12-17**
Some species of plants will only grow in wetlands. *Courtesy of James Strawser, The University of Georgia.*

most of the growing season. If wetlands are destroyed, many of these plants may be lost forever.

The EPA has established regulations that protect wetlands by restricting their use. Incentives are offered to landowners to promote the preservation of wetland areas. Permits must be obtained from the government when wetlands are to be disturbed. Through these measures valuable wetlands will be protected and in turn the whole environment will be enhanced.

## Summary

Agriculturists will continue to help protect the environment. It is in their best interest and in the public interest. We must have food and fiber. This can only come about through agriculture. Scientists and other concerned individuals will continue to find new and better ways of producing food and fiber and of protecting the environment.

# 12 CHAPTER REVIEW

## Student Learning Activities

1. Survey your community and decide if the environment in your area is being cared for. If so, list the measures taken to preserve the environment. If you feel there is room for improvement, outline steps that can be taken to solve the problems.

2. Locate an area where there has been soil erosion. Outline a plan that can help prevent more loss of the soil. Be sure to list the major factors that contributed to the erosion.

3. Go to the library and locate a story of a pollution problem that was solved. Report to the class on your findings.

## True/False

1. ___ The deeper the water supply and the heavier the clay content of the soil, the more likely the water will be polluted.

2. ___ The National Environmental Policy Act led to the establishment of the Environmental Protection Agency.

3. ___ Since 1970, a lot of progress has been made in cleaning the air and water.

4. ___ Erosion is the wearing away of soil by wind or water.

5. ___ Research indicates that the sediments from soil cause approximately $6 billion in damage per year.

6. ___ Diversion ditches are ponds of water where waste is directed.

7. ___ Over half of the wetlands in the United States have been destroyed for construction or agricultural purposes.

8. ___ The Soil Conservation Service is concerned with minimizing pesticides in soil.

9. ___ All crops are now grown with extensive plowing.

10. ___ Nitrates cause plant life in streams and lakes to flourish.

# Multiple Choice

1. One of the most precious natural resources we have is
   a. crops.
   b. soil.
   c. trees.

2. Rainfall replenishes two types of water storage: groundwater and
   a. surface water.
   b. lagoons.
   c. stored water.

3. The water in streams, lakes, and oceans is
   a. groundwater.
   b. surface water.
   c. polluted water.

4. Formations in sand, rock, or gravel that hold water are
   a. wells.
   b. wetlands.
   c. aquifers.

5. Pollution comes about as a result of contaminants entering the water in two main ways:
   a. point and nonpoint sources.
   b. rain and snow.
   c. fish and birds.

6. The largest origin of nonpoint pollution in the United States is
   a. rain.
   b. industry.
   c. agriculture.

7. Our largest environmental problem stemming from agriculture is that of
   a. soil erosion.
   b. dairy farming.
   c. excess pesticides.

8. Permits must be obtained from the government when these are disturbed:
   a. wells.
   b. wetlands.
   c. aquifers.

9. Years of rigorous testing are required to ensure the safety of
   a. crops.
   b. wells.
   c. pesticides.

10. Resources that can be replaced are
    a. renewable.
    b. residual.
    c. hydrological.

# Discussion

1. What are natural resources?

2. What are renewable resources?

3. Why must farming practices become more intense?

4. What is the difference between modern and older pesticides?

5. What are some uses for water?

6. Where does agricultural pollution come from?

7. What is soil erosion?

8. Why is sediment pollution such a large problem?

9. Why is it better that crops are grown with a bare minimum of plowing?

10. Why are wetlands so important?

# The Livestock Industry

## Student Objectives

*When you have finished studying this chapter, you should be able to:*

- Discuss the extent of the U.S. meat industry.

- Discuss the different phases of the beef industry.

- Explain how pork is produced in the United States.

- Discuss the extent of the sheep industry.

- Describe the types of and uses for horses in this country.

- Describe the small animal industry.

The livestock industry in the United States is a tremendously large industry. Animals are raised in all 50 states. They are used for meat, milk, and dairy products (see Chapter 14), clothing (wool and leather), and for recreation (pets and horses). For as long as there has been a country, Americans have raised large numbers of livestock. Our country is well suited for growing livestock. Much of the open land in this country that is not suitable for growing row crops is suitable for growing animals.

## The Meat Industry

The average person in this country consumes 48 pounds of **beef** and veal, 29 pounds of **pork,** and 24 pounds of **poultry** each year. Very few nations in the world even come close to us in the **per capita consumption** of meat.

Some health authorities have expressed concern over the consumption of so much meat. Research has shown that a substance called cholesterol in the fat of meat can cause heart problems. Most health experts agree that a moderate intake of meat is necessary for a healthy diet. Meat is among the most nutritionally complete foods that humans consume. Foods from animals supply about 88 percent of the vitamin $B_{12}$ in our diets because this nutrient is very difficult to obtain from plant sources. In addition, meats and animal products provide 67 percent of the riboflavin, 65 percent of the protein and phosphorus, 57 percent of the vitamin $B_6$, 48 percent of the fat, 43 percent of the niacin, 42 percent of the

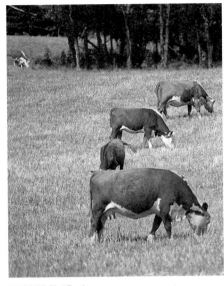

**FIGURE 13-1**
Much of the land in the United States that is not suitable for row cropping is suitable for growing animals. *Courtesy of James Strawser, The University of Georgia.*

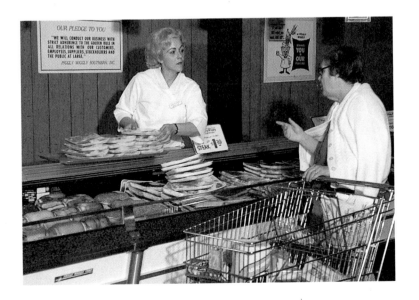

**FIGURE 13-2**
Very few nations in the world eat as much meat as Americans. *Courtesy of James Strawser, The University of Georgia.*

vitamin A, 37 percent of the iron, 36 percent of the thiamin, and 35 percent of the magnesium in our diets. Problems with cholesterol can be avoided by exercise and eating a reasonable amount of meat.

## The Beef Industry

Beef is by far the largest segment of the meat industry. In the United States, there are over 800,000 producers of beef. Over three-quarters of all of the money earned from the meat industry comes from beef. The average size of the beef herds in the United States is about 100 animals.

In the United States there are over 40 different breeds, and many different combinations or crosses of these breeds. A **breed** is a group of animals with a common ancestry and like characteristics. The offspring of these animals look like their parents. Livestock producers choose breeds based on several factors. The market where the animals will be sold, the environmental conditions the animals will be produced in, and the personal likes and dislikes of the individual producer may be considered.

Some breeds are large and produce a large **carcass,** some mature at a smaller size and produce

**FIGURE 13-3**
A breed is a group of animals with a common ancestry and similar characteristics. *Courtesy of James Strawser, The University of Georgia.*

a smaller carcass. A carcass is that part of a meat animal remaining after slaughter and the hide, head, feet, and internal organs have been removed. Both sizes have a place in the market and both are produced. Some breeds are adapted to hot, humid climates and some tolerate the cold and snow better. A producer may like the color pattern or the docile nature of a particular breed and prefer to produce that breed.

Some breeds make excellent mothers while other breeds grow rapidly and produce high-quality, meaty carcasses. Because of this, some breeds are referred to as sire breeds and some are referred to as dam breeds. A sire is the father and the dam is the mother.

Most animals produced for slaughter are **crossbreeds.** A crossbreed is an animal with parents of different breeds. A crossbreeding program helps producers to take advantage of the good points of both types of animals.

The beef industry has four major segments: purebred operations, cow-calf operations, stocker operations, and feed lot operations. **Purebred** cattle are produced in the first phase of the industry. The term purebred means that the animal is of a cer-

**FIGURE 13-4**
A crossbreed is an animal with parents of different breeds. *Courtesy of Cooperative Extension Service, The University of Georgia.*

tain breed and has only that breed in its ancestry. Purebred animals are eligible for registration, which certifies that the animal is purebred. The purpose is to produce what is known as seed stock cattle. These cattle are used as the dams and sires of calves that will be grown out for market. These are usually those that are of the sire type or dam type of animal.

The second phase is the cow-calf operations, where the calves are produced. The calves are weaned from their mothers and sold to the next phase of the industry. Most of these calves are crossbred animals from purebred parents of different breeds. A large part of this industry is centered in the southern and western states. The mild winters of the South are ideal for calving in the winter. In the West, producers can take advantage of the vast amounts of government lands that are open to grazing for a small fee. Often, cows are left on free range (not fenced in) to have their calves and are then rounded up, weaned, and sold.

Calves are usually sold upon weaning. They are weaned at about six to seven months in age. They generally weigh approximately 500 to 600 pounds. At this age and weight they are ready to begin feeding for market.

The next phase of the industry is stocker operations, a step between the weaning of the calves and the finishing or fattening of the animals prior to slaughter. The stocker purchases the animals from the cow-calf producer and sells them to the feedlot operator. The stocker provides a transition period for the calves between the time they are weaned and before they are put in the feedlot. The calves are fed a **ration**, which is the amount of food necessary during a 24-hour period, that ensures that they make sufficient gains to be placed in the feedlot.

It is not uncommon for feedlot owners to also run stocker operations. Such arrangements are eco-

**FIGURE 13-5**
The purpose of a cow-calf operation is to produce calves to be grown out for market. *Courtesy of James Strawser, The University of Georgia.*

FIGURE 13-6
Stocker operations provide a step between weaning and the feedlot. *Courtesy of Ronnie Silcox.*

nomical in that fewer transportation costs are incurred if the two types of operations are close together. The trend in the industry has been away from the stocker industry as new research reveals production methods that allow cows to wean heavier calves.

The feedlot operation is the final phase before the animals are sent to slaughter. Here the animals are fed on a high concentrate ration designed to put on the proper amount of fat cover. A concentrate is a feed that is high in grain content. Grain contains a lot of carbohydrates and helps an animal

FIGURE 13-7
Cattle in a feedlot are fed a high concentrate ration. *Courtesy of American Limousin Association.*

to fatten. A roughage is a feed such as grass or hay that is high in fiber. The producers usually want their animals to be marketed when the cattle reach a sufficient fat cover to allow the animals to grade "low choice."

Many feedlots are situated in the Midwest because this section of the country produces the most grain. It is usually more economical to feed the animals there rather than ship the grain across country. An exception is the state of Texas, which has more feedlots than any other state. When the animals are slaughtered, they are generally around 18 to 24 months in age and can weigh from 800 to 1,500 pounds. This age and size offer consumers the type of beef they prefer.

## The Pork Industry

As producers and consumers of pork, the United States ranks behind Asia and Europe. Every year Americans produce over 85 million head of hogs. In terms of meat, pork production and consumption ranks second only to beef in this country. Pork is distributed throughout the country, although some groups, such as Moslems and Jews, do not eat pork for religious reasons.

Much of the pork produced in this country comes from the midwestern states of Iowa, Illinois, Indiana, Minnesota, and Nebraska. These are the states that produce corn, the major grain fed to swine. Also there are large numbers of pigs in the South, where mild winters help lower the cost of production.

There are not as many breeds of hogs as there are breeds of cattle grown in this country. Breeds are categorized as mother or sire breeds. The mother breeds include Landrace and Yorkshire, which have a large number of pigs per litter and produce lots of milk for their young. The sire breeds, such as the

**FIGURE 13-8**
Pork production and consumption ranks second only to beef in the United States.

FIGURE 13-9
A Landrace female makes a good mother and is considered a dam breed. *Courtesy of the American Yorkshire Club.*

FIGURE 13-10
Pigs are generally born in a furrowing crate to protect them from being crushed. *Courtesy of University of Minnesota Agricultural Experiment Station.*

Duroc and the Hampshire, characteristically grow rapidly and produce well-muscled, meaty carcasses.

Most of the market hogs produced are crosses of the various breeds. For example, a Yorkshire female mated to a Duroc male provides an excellent cross. The Yorkshire sow provides a large litter and plenty of milk and the Duroc boar sires pigs that are vigorous and fast growing. In addition, crossbred pigs are healthier and do better than purebred pigs.

Two phases of the industry are the farrowing and finishing operations. The two phases can be separate or they can be operated together. Some producers prefer to raise only feeder pigs (pigs that are sold shortly after weaning) and some prefer to buy feeder pigs and finish them as their only operation.

Most pigs are farrowed in climate-controlled houses where the mother is kept in a crate to prevent her from injuring the pigs as she lies down. Good producers make quite an effort to provide an environment that is clean, dry, and comfortable

# Career Development Events

**I**f you have an interest in livestock and would like to work in the livestock industry, you could benefit from participating in several FFA career development events. The FFA Livestock Judging event is for teams consisting of four members. Each member competes individually and the final team score is calculated by adding the three highest individual scores. The FFA livestock judging event aids classroom instruction by teaching students practical skills in selecting quality livestock. All livestock producers know that the selection of quality livestock is essential to be competitive and profitable in the industry. Some skills and talents you will need to participate in the event are understanding the breed characteristics of beef, swine, and sheep. The ability to rank classes of animals and give reasons for placement of animals is also required. Team members independently evaluate and place each livestock class. After the judging, students are asked to give reasons for the rationale they used to place the livestock. Points are awarded for each class placing and correct reasons. This FFA event is usually held at the chapter, area or district, state and national levels.

Another FFA team event related to the livestock industries is the Meats Judging Career Development Event. Like the livestock judging event, teams are made up of four FFA members. As a participant in this event, you will gain valuable knowledge of meat identification and grading. This knowledge will benefit you in the future if seeking a job in the meat packing industry or if you are a consumer of meat. Preparation for this event combines team practice and individual study. The team should practice as often as possible. In most career development events, students identify primal cuts of beef and pork, retail cuts of pork, beef and lamb, and quality or yield grades of carcasses. By visiting meat processing plants and local grocery meat departments, the identification of wholesale and retail cuts can be reviewed. Each contestant will need to study for the written test relating to meat selection, storage, cookery, nutrition, and safety. Situational problems involving the least-cost formation of a batch of meat products such as hamburger or bologna are given. The student works through procedural questions and the actual determination of the least-cost price. The FFA Meats Judging event develops employment knowledge with practical skill applications preparing students for entry-level employment in the meat packing industry.

Students may compete in individual events in this area through the FFA Proficiency awards. All FFA members who complete Supervised Agricultural Experience Programs (SAEPs) as part of their agriculture education instruction are eligible for the award. Livestock proficiency awards are available in beef production, dairy production, diversified livestock production, and sheep and swine production. Proficiency Award medals and certificates are provided by the National FFA Foundation. Chapter winners may compete at the state level where the winner earns a cash award and a plaque plus travel expenses to the National FFA Convention. The top FFA member in the nation in each award area is recognized and awarded at the National Convention.

In the Meats Judging Career Development Event, students test their knowledge of meat quality and yield grades as well as the identification of different cuts of meat. *Courtesy of Carol Duval, National FFA Organization.*

FIGURE 13-11
Hogs are grown to market weight on a finishing floor. *Courtesy of James Strawser, The University of Georgia.*

for both the mother and the piglets. The pigs are generally weaned from the mother at six weeks, although they may be as young as three weeks or as old as eight weeks.

After weaning, the pigs are placed together with pigs of similar age and size in what is called a confinement operation. This means that the hogs are kept in a pen together rather than running loose on a pasture. Sufficient space is allowed for the pigs to be comfortable and to grow at a fast rate. The pigs should be finished (reach the proper market weight and condition) at about 20 weeks. Packers like to buy market hogs that weigh in the range of 220 to 260 pounds.

## The Sheep Industry

Compared to beef and pork, Americans eat relatively little lamb and mutton. Lamb refers to meat from a sheep that is less than a year old; mutton refers to meat from a sheep that is over a year old. In many parts of the world, lamb and mutton are a basic part of the diet. The U.S. per capita consumption of mutton and lamb is only about 2.5 pounds. Of this, about 95 percent is lamb and only about 5

**FIGURE 13-12**
Annual consumption of lamb is only about 2.5 pounds per capita in the United States. *Courtesy of James Strawser, The University of Georgia.*

percent is mutton. Americans seem to have never developed a taste for the stronger flavored mutton.

One advantage of producing lambs for market is that very good quality lambs can be produced on grass without having to be fed a lot of expensive grain. Although an increasing number of lambs are being fed on grain in feedlots, roughages still make up about 90 percent of all the feed consumed by sheep.

The leading states in sheep production are Texas, California, Wyoming, Colorado, South Dakota, Montana, New Mexico, Utah, and Oregon. Sheep can make better use of lower-quality forage than can cattle. For this reason they can be successfully grazed on poorer grazing lands of the desert areas of the West. In addition, a drier climate helps reduce parasite and disease problems associated with sheep that are grown in the more humid areas.

Sheep breeds are generally grouped according to the type of wool the animals grow. Wool type is broadly classified as fine wool, long wool, medium wool, hair, and fur. The medium wool breeds are used most often to produce lambs for slaughter. Medium wool breeds commonly used to produce slaughter lambs are Suffolk, Hampshire, South Down, and Dorset.

**FIGURE 13-13**
The Dorset sheep is an example of a medium wool breed of sheep. *Courtesy of James Strawser, The University of Georgia.*

**FIGURE 13-14**
Much of our history has been built with the power of horses and mules.
*Courtesy of John Deere & Company.*

# The Horse Industry

Humans have used horses for transportation, work, and war as far back as recorded history goes. Almost all civilizations at one time relied on horses or donkeys. In this country much of our history has been built around power supplied by horses and mules.

The number of horses and mules in this country grew until the 1920s when the car, truck, and tractor caused a sharp decline in their numbers. From that time until around 1960, the numbers steadily declined. Since the 1960s the number of both horses and mules has increased dramatically. While some horses and mules in the United States are used for work, most of the animals are used for recreational purposes.

In modern times horses are generally categorized as light horses, draft horses, or ponies. Light horses are animals that weigh 900 to 1,400 pounds. These horses are further divided according to use. Gaited saddle horses and walkers are used primarily for pleasure riding and show. They have different ways in which they walk or trot that seem graceful and flowing. Driving horses are used to pull carriages that are either four or two wheel and are used for recreation or for show.

Stock horses are used all across the country to work cattle or other animals. Race horses are used to compete with each other as a spectator sport.

Draft horses are those breeds that weigh over 1,400 pounds. At one time these animals provided the power for pulling heavy loads such as wagons, plows, and other agricultural implements. Today these animals are used in pulling competitions, shows, and parades.

Ponies are breeds of horses that weigh 500 to 900 pounds when mature. While these are used to pull carriages and for show, the majority of ponies are used as horses for children.

Altogether the horse industry in this country is a $15 billion industry. Surprisingly, horse racing ranks third behind baseball and auto racing among spectator sports in the United States. Each year over 75 million people attend thoroughbred and harness racing. Each year there are around 7,000 horse shows in this country where young people and adults compete in a variety of different events.

**FIGURE 13-15**
Today most horses are grown for recreation. *Courtesy of Michael Dzaman.*

**FIGURE 13-16**
Ponies are breeds of horses that weigh between 500 and 700 pounds at maturity. *Courtesy of Multi-World Champion Shetland Harness Pony.*

Many horses are owned as individual saddle horses that are used as recreation and are never raced or shown.

## Small Animals

A growing industry in the United States is that of producing small animals. These animals range from rabbits raised for meat to animals such as gerbils and mice raised for pets. Animals such as guinea pigs and rats are raised for research. These animals are used to test everything from drugs to toothpaste.

### Rabbit Production

People have raised rabbits for food for hundreds of years. Although most rabbits are produced by part-time growers, there are several large commercial operations in this country. Americans consume around 10 to 13 million pounds of rabbit meat per year. Some of the meat consumed in this country is imported from Europe where the rabbit industry is larger. France is the leading producer.

**FIGURE 13-17**
Rabbits are raised in cages called hutches. *Courtesy of Calvin Alford, The University of Georgia.*

There are several advantages to raising rabbits. Rabbits are easily raised under almost any climatic condition. Most are raised indoors in cages called hutches that consist of woven wire with boxes for the rabbits to sleep in and to have their young.

Many restaurants now offer several dishes that are prepared using rabbit meat. The USDA points out that rabbit meat is one of the most nutritious meats available. Not only is it high in protein and low in fat and cholesterol, but it is easily digested and is very flavorful. In addition to the meat, rabbits are used for several purposes. The fur is used in making coats, hats, liners for boots, and for toys. Scientists use the animals in medical research and other areas. Many animals are sold as pets because of their docile nature, clean habits, and cuddly fur.

Rabbits can be a relatively inexpensive source of meat, but Americans do not consume as much rabbit meat as other peoples of the world. While we may accept the production and slaughter of cattle, hogs, and chickens, the thought of slaughtering cute, cuddly animals such as rabbits is repulsive to many.

## Summary

The livestock industry in this country is broad and diverse. Most experts say that in the future it will become even more diverse. In the future production will expand to include some exotic animals. Such animals as ostriches, emus, and alligators may be grown for food. This may seem far-fetched, but remember that at one time cattle, horses, and sheep were wild animals before humans tamed them.

**FIGURE 13-18**
In the future, animal agriculture may expand to include animals that are now considered to be exotic. *Courtesy of James Strawser, The University of Georgia.*

# 13 CHAPTER REVIEW

## Student Learning Activities

1. Make a list of all the products in your home that come from animals. Determine what type of animal was used for the product. For example, a leather belt was probably from a beef animal and a wool sweater was probably from a sheep.

2. Report on the most widely grown agricultural animal in your state. List the reasons why this animal is popular in your state.

3. Design a plan for growing animals that are not grown in your area. Give your reasons for growing the animals, why your area would be suitable, and how you would market the animals.

## True/False

1. ___ The livestock industry in the United States is a tremendously large industry.

2. ___ Our country is not suited for growing livestock.

3. ___ Meat is among the most nutritionally complete foods that humans consume.

4. ___ The average size of the beef herds in the United States is about 50 animals.

5. ___ Most feedlots in this country are situated in the South.

6. ___ Ostriches, emus, and alligators may be grown for food.

7. ___ Calves are usually sold upon weaning.

8. ___ The United States leads the world in the production and consumption of pork.

9. ___ Americans eat more mutton than lamb.

10. ___ Sheep can make better use of lower quality forage than can cattle.

# Multiple Choice

1. The largest segment of the meat industry is
   a. beef.
   b. poultry.
   c. fish.

2. About 90 percent of all feed consumed by sheep is
   a. mutton.
   b. corn.
   c. roughage.

3. The major grain fed to swine is
   a. corn.
   b. wheat.
   c. roughage.

4. Horses that weigh 900 to 1,400 pounds are called
   a. light.
   b. purebred.
   c. crossbreed.

5. The third largest spectator sport in the United States is
   a. horse racing.
   b. football.
   c. auto racing.

6. Most rabbits are raised indoors in cages called
   a. hutches.
   b. sires.
   c. dams.

7. A substance found in meat that can cause heart problems is called
   a. mutton.
   b. roughage.
   c. cholesterol.

8. Over three-quarters of all of the money earned from the meat industry comes from
    a. chicken.
    b. beef.
    c. turkey.

9. The mild winters of the South are ideal for
    a. calving.
    b. racing.
    c. wool production.

10. The step between the weaning of the calves and the finishing or fattening of animals prior to slaughter is
    a. registration.
    b. feedlot operations.
    c. stocker operations.

## Discussion

1. List some uses for animals.

2. Why are health authorities expressing concern over the consumption of so much meat in the United States?

3. What vitamins do meat and animal products provide?

4. How do livestock producers choose the breed they grow?

5. What are the four major segments of the beef industry?

6. What is an advantage of producing lambs?

7. Why is it better to raise sheep in a drier climate?

8. How is wool classified?

9. Why did the number of horses and mules used in this country decline?

10. What are the advantages of raising rabbits?

# The Dairy Industry

KEY TERMS

dairy products

selective breeding

BST

fermentation

milkfat

homogenization

pasteurization

cheese

yogurt

## Student Objectives

**When you have finished studying this chapter, you should be able to:**

- Explain the importance of the dairy industry.
- Discuss the controversy surrounding the use of BST.
- Tell how milk is produced.
- Explain how milk is processed and graded.
- Discuss the cheese making process.
- Tell how yogurt is made.

A large portion of agriculture is the dairy industry. This industry includes all of the products made from milk. Around 13 percent of all sales of farm commodities are from **dairy products.** Americans use a lot of dairy products. On average, we consume 28.8 gallons of milk, 23.7 pounds of cheese, 16 pounds of ice cream, 4.3 pounds of butter, and 4.3 pounds of yogurt per person, per year. These products are high in protein and contain essential minerals and vitamins. Milk has been described as nature's most perfect food. It is the only

**FIGURE 14-1**
Around 13 percent of all sales of farm commodities are from dairy products. *Courtesy of James Strawser, The University of Georgia.*

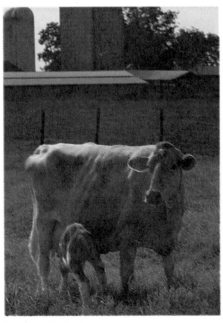

**FIGURE 14-2**
Milk is intended by nature to be a food for young animals. *Courtesy of James Strawser, The University of Georgia.*

substance intended by nature for no other purpose than as food. Milk is obtained from cows that produce it to feed their calves. Since it is the only food a very young animal receives, it must be filled with nutrients. Humans make use of this nutritious food.

**Selective breeding** of dairy cattle has developed cows that produce much more milk each year than a calf could consume. This allows us to use the surplus for our food. There are about half the number of dairy cows in this country than there were in 1950. However, the amount of milk produced remains about the same. This means that modern dairy animals produce twice the milk their ancestors did.

In recent years consumers have been concerned over a substance given to dairy cattle to increase production. This substance, called bovine somatotropin or **BST,** is a naturally occurring hormone that helps to stimulate milk production. Scientists have known for many years that cows that get additional amounts of BST significantly increase their milk production. Until recently, this substance was scarce and very expensive. However, due to genetic engineering, the hormone can now be produced quickly and cheaply. The use of BST has brought about controversy over the safety of the substance. Some say that no studies have been conducted to determine the long-term effects of drinking milk from cows that have been given BST. They point to the fact that large doses can cause inflammation of the cows' udders and that this may be a sign that the hormone is not safe. Proponents point out that the cows are only given very small doses and that no ill effects have been discovered. Also, the National Institute of Health and the Food and Drug Administration have declared BST to be safe for both cows and humans who consume the milk. All milk contains a small amount of BST. BST is a naturally occurring substance. Milk from cows given BST has only a minute amount more BST than milk from cows that have had no artificial BST. All BST in the milk is digested and none gets into the bloodstream of humans.

**FIGURE 14-3**
Most dairy animals in the United States are of the Holstein breed. *Courtesy of James Strawser, The University of Georgia.*

## Milk Production

Most of the dairy cattle in the United States are Holstein. This breed produces more milk than most other breeds. Some breeds, like the Jersey, produce milk with a higher fat content. This was once a desirable characteristic. Today, however, consumers want foods with less fat content. Most of the milk produced in the United States comes from Wisconsin, California, New York, Minnesota, and Pennsylvania. These leading states have the climate necessary for cows to do well and large cities where the milk is marketed. However, every state produces and processes some milk.

Most dairy animals are fed on pastures. Grass makes a relatively inexpensive feed for the cows. Some dairies now feed their cows in lots rather than pastures. These cows are fed on a feed called silage. This feed is made by chopping green corn (ears stalks and leaves). The chopped corn goes through

**FIGURE 14-4**
Dairy cows are fed on silage, which is made from chopped corn. *Courtesy of James Strawser, The University of Georgia.*

a process known as **fermentation,** which is a chemical change, that helps to preserve the feed.

In order for cows to produce milk they must have a calf each year. Remember that the function of milk is to feed the young animal. The cycle of birth stimulates the mammary system of cows to produce milk. Most of the cows are bred artificially. This offers several advantages. By using artificial insemination (AI) producers do not have to go to the trouble and expense to keep a bull. Also, problems with sexually transmitted diseases are eliminated. AI gives the advantage of being able to select semen from a wide variety of superior bulls at a relatively low price. Daughters from these bulls can be used for replacement heifers and the quality of the herd can be raised.

When the calf is born, the first milk that is produced is different. This milk, called colostrum, is loaded with the mother's antibodies. These help the young animal to ward off disease during the first few days of life. Also, colostrum contains substances that help to open up the calf's digestive system and get it to function properly. Although this milk is essential to the young animal, it is considered unfit for human consumption. Cows that have

**FIGURE 14-5**
The first milk a newborn calf gets is called colostrum. *Courtesy of James Strawser, The University of Georgia.*

just given birth are separated and the colostrum is not mixed with milk intended for the market.

The cows are milked in an area called a parlor. These areas are scientifically designed so the cows can be handled gently. They come into the parlor and enter a station where they obtain feed. The cow stands still while she is being milked. A milker enters the parlor and milks a small amount of milk by hand into a container called a strip cup. This gives the milker an opportunity to check for problems in the milk. Lumps or blood in the milk could indicate a disease called mastitis. This disease is caused by injury to the cow's udder. Milk from cows with mastitis is not used for human consumption. Milking and discarding the first two or three squirts of milk also helps to cleanse the teat.

The udder is then washed using a warm water solution and dried. The teat cups are then attached and the milking begins. These cups are lined with a soft material that is attached to a tube that gently pulls a vacuum on the cup to draw milk from the teat. The teat cup fits snugly on the cow's teat and pulsates by means of a vacuum on the lining of the cup. The milk is all removed in three to six minutes depending on the individual cow and the amount of milk she gives. Care is given that the cups are left on the proper amount of time. Too little time and the udder will not be milked out, and too much time and injury to the udder can result. The teats are then treated with a disinfectant and the cows are released. The teat cups are cleaned between cows to prevent the spread of disease.

The milk is drawn through lines and into a holding tank. In modern dairies the cooling of the milk begins in the lines. The lines are refrigerated so that the milk is cooled as it is transported to the tank. Here it is rapidly cooled to around 40 °F to prevent the multiplication of bacteria and to prevent the milk from spoiling. After all the cows have been

FIGURE 14-6
A milker gently attaches teat cups that draw out the milk. *Courtesy of James Strawser, The University of Georgia.*

milked, the lines, teat cups and all of the equipment are thoroughly cleaned. About every other day the milk is picked up by a tanker truck and hauled to the processing plant. At the plant, the milk is tested

## Career Development Events

**FFA**

The FFA offers many exciting and challenging experiences. Members who participate in team events spend many hours learning the basics in the classroom and practicing skills after school. Dairy Cattle Judging and Milk Quality and Dairy Food competition are two events that require cooperative efforts of team members. If you have an interest in dairy products, plan to own dairy cattle, have a Supervised Agricultural Experience Program (SAEP) in the dairy industry, or work on a dairy farm, you would be a good team member for these events.

The Dairy Cattle Judging Career Development Event trains students in the selection of a quality dairy herd. Practice and experience are the only ways to develop the skills needed to select quality animals. A well-prepared team should understand what a quality dairy animal looks like and be familiar with dairy herd improvement records. Most Dairy Judging teams consist of four FFA members. Usually through a competitive process, the agriculture teacher selects a team to represent the FFA chapter in competition. The FFA Dairy Judging event requires each team member to know the breeds of dairy cattle and traits that yield high milk production. During the event each member is responsible for evaluating and placing groups of animals in numerical order. The student must also give reasons for each animal's placement. After each team member has evaluated and placed all dairy groups, individual scores are calculated. The top three scores are added to determine the total team score and placement. All state winning teams advance to the National Dairy Judging Career Development Event.

Another FFA event that helps prepare students for careers in the dairy industry is the Milk Quality and Dairy Foods Event. Hundreds of products are manufactured by the dairy industry. Every dairy product is graded by its quality. It is important that workers in the dairy industry and consumers know what makes a high-quality product.

for the number of bacteria, drug residue, and any indications of infection. Milk with high levels of bacteria, drug residues, or signs of infection is discarded.

To participate in this event, a student must have a keen sense of sight, smell and taste to distinguish grades of dairy products. Classroom study of milk content and flavor, classification of cheeses, and the parts and operation of milking equipment is essential. The skills required to rank dairy products in order of quality can only be developed by practice. The Milk Quality and Dairy Foods Event is a great opportunity to prepare FFA members with skills required to be competitive dairy workers and knowledgeable consumers.

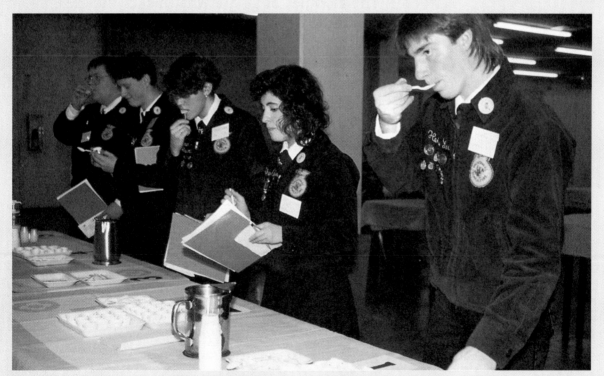

Students competing in the Milk Quality and Dairy Foods Career Development Event must distinguish between grades of dairy products. *Courtesy of Carol Duval, National FFA Organization.*

When the milk arrives at the processing plant it is thoroughly filtered to remove any foreign particles. The milk is allowed to sit so the cream may be removed from milk to be sold as low fat milk. Cream is the part of the milk that contains fat. As consumers are becoming more conscious of the amount of fat in their diet, they want milk that is lower in **milkfat** than whole milk. In recent years sales of lowfat and skim milk have increased sharply. Lowfat milk is milk that has had the percentage of milkfat lowered to between 0.5 and 2 percent. Skim or nonfat milk is a milk containing less than 0.5 percent milkfat. The milkfat that is removed from the milk is then used to make ice cream and other cream products.

Whole milk contains around 4 percent milkfat. The globules of fat are what makes the cream float to the top of raw, unprocessed milk. In a process called **homogenization,** the large cream globules are reduced in size to the size of the milk globules. The processed milk, called homogenized milk, will not separate out.

To kill any harmful organisms in the milk, the milk is first heated and then cooled in a process called **pasteurization.** In pasteurization, the

**FIGURE 14-7**
Milk is sold as whole, lowfat, and skim milk. *Courtesy of James Strawser, The University of Georgia.*

**FIGURE 14-8**
In the process of pasteurization, harmful bacteria are destroyed. *Courtesy of James Strawser, The University of Georgia.*

temperature of the milk is raised and then is promptly cooled. This kills harmful bacteria that might cause illness in humans. The time and temperature must be precisely controlled to protect the nutritive value and flavor of the milk.

Milk is graded according to the dairy from which it came. Dairies that sell grade A milk must pass rigid standards that include cleanliness and other conditions. Only Grade A milk can be used as fluid or beverage milk. Grade B milk can only be used for manufactured dairy products. Since the production of Grade A milk far exceeds the demand for fluid milk, Grade A is used in processing also. For pricing purposes, the milk is classified as Class I, II, or III. Class I is used for beverage consumption; Class II is used for manufacturing soft products such as ice cream, yogurt, and cottage cheese; Class III is used with Grade B milk in the processing of cheese, butter, and nonfat dry milk.

## Cheese

The United States produces more **cheese** than any other country in the world. There are literally hundreds of types of cheese. They are broadly classified according to the type of milk used and the type of processing. Cheeses are often named for the town or region where they originated. Most of the cheese produced in the United States uses cow's milk. In some parts of the world goat's milk (called Chevre cheese) or sheep's milk (called Roquefort) may be used in the process. Other cultures may make cheese from the milk of horses or camels.

Cheese contains one of the highest concentrations of nutrients of any food we eat. Only 3.5 ounces supplies about 36 percent of the protein, 80 percent of the calcium, and 34 percent of the fat recommended in our daily diet. It is rich in essential amino acids, minerals, vitamins, and has a high caloric content.

Most cheese begins with milk that has been pasteurized to help prevent the multiplication of harmful bacteria. The milk is placed in a large vat and a bacteria culture is added. This culture is called a starter culture because it starts the process of fermentation, which changes sugars to acids. These acids cause the proteins in the milk to coagulate, that is, to form into a solid. To further the process an enzyme called rennet is added. An enzyme is a substance that speeds up or stimulates a chemical process. Rennet is obtained from the stomachs of calves. During this step, large paddles turn the milk to ensure that the starter bacteria and the rennet are evenly distributed.

The solid resulting from this step is called curd. The liquid is drained off and is called whey. Whey is used in processing human food and manufacturing animal feed. The curd is cut into small cubes to increase the surface area to allow the drainage of the whey. After the whey is drained off, the curd sits until it forms a solid mass. The curd is then heated to cause the curd to contract and further expel the whey. The amount of heat and length of time the cheese is heated depends on the type of cheese

**FIGURE 14-9**
To begin the cheese making process, milk is placed in a large vat and starter cultures of bacteria are added. *Courtesy of Tillomook County Dairy.*

**FIGURE 14-10**
The solid is called curd and the liquid that is drained away is called whey. *Courtesy of Tillomook County Dairy.*

**FIGURE 14-11**
Cheese is coated, packaged, and stored for aging. *Courtesy of Tillomook County Dairy.*

being made. The cheese is salted and pressed into a metal form or a cloth bag.

The final step in cheese making is the curing or ripening of the cheese. The cheese is placed in an environment that is controlled for temperature and humidity. The specific conditions vary with the type of cheese being made. During this time, enzymes produced from the starter bacteria change the flavor, texture, and appearance of the cheese. The cheese is packaged in a coating of paraffin or is wrapped in cloth or plastic. It is allowed to age a specific amount of time before it is sold.

# Yogurt

**Yogurt** is also processed from milk. This food product has been around for hundreds of years. It has gained in popularity during the past few years. When frozen, it is delicious and is lower in calories than ice cream. Many people prefer frozen yogurt to ice cream. Yogurt comes in a wide variety of flavors and is often flavored with fruit.

Processing begins by heating concentrated milk. It is cooled rapidly and cultures of bacteria are added. These cultures cause the milk to ferment. In this process the required acidity and the yogurt flavor are produced. The acidity helps to give the yogurt a tart flavor and also helps to preserve it. Before it is packaged fruit or other flavorings may be added.

## Summary

Dairy is a very significant part of our agricultural industry. It provides many jobs in the production, processing, and distribution of milk and milk products. It also provides Americans with wholesome, nutritious, and good tasting food. In the future new uses may be found for milk products that will make the industry even more important.

# 14 CHAPTER REVIEW

## Student Learning Activities

1. Collect all the information you can about the use of BST with dairy cattle. Decide whether or not you believe the product should be used. Defend your position to the class.
2. Visit a large grocery store and make a list of all the different types of dairy products that are available. Interview the manager to find out which sell the best.

3. Choose a particular type of cheese and write an essay on the cheese. Include how the cheese is different from other cheese, where and how it originated, where it is made, and the process used in the manufacture.

# True/False

1. ___ Around 50 percent of all sales of farm commodities are from dairy products.

2. ___ Milk has been described as nature's most perfect food.

3. ___ There are about half the number of dairy cows in this country than there were in 1950.

4. ___ BST is a hormone that is very expensive to produce.

5. ___ All milk contains a small amount of BST.

6. ___ Only a few states process and produce milk.

7. ___ In order for cows to produce milk they must have a calf each year.

8. ___ Colostrum is not used for human consumption.

9. ___ The United States produces more cheese than any other country in the world.

10. ___ Milk is the only substance that is intended by nature for no other use than as food.

# Multiple Choice

1. A very large portion of animal agriculture is the
   a. dairy industry.
   b. pesticide industry.
   c. wheat industry.

2. The dairy industry includes all of the products made from
   a. cows.
   b. colostrum.
   c. milk.

**3.** Dairy products are high in
   a. hormones.
   b. mastitis.
   c. protein.

**4.** Most of the dairy cattle in the United States are
   a. Holstein.
   b. Angus.
   c. Sires.

**5.** A relatively inexpensive feed for cows is
   a. wheat.
   b. grass.
   c. grain.

**6.** Most of the cheese produced in the United States comes from
   a. goats.
   b. cows.
   c. artificial dairies.

**7.** The enzyme rennet is obtained from calves'
   a. udders.
   b. kidneys.
   c. stomachs.

**8.** Colostrum is loaded with
   a. chemicals.
   b. antibodies.
   c. acids.

**9.** Mastitis is caused by injury to the cow's
   a. stomach.
   b. udder.
   c. kidneys.

**10.** Whole milk contains about 4 percent
   a. colostrum.
   b. calcium.
   c. milkfat.

# Discussion

1. What are the advantages of artificial breeding?

2. Why is colostrum good for newborn calves?

3. How is milk classified?

4. What animals are used for their milk?

5. What nutrients are supplied by cheese?

6. What purpose does acidity have in yogurt?

7. Why is the dairy industry so important?

8. What does fermentation do to milk?

9. Why are bacteria added to milk to make yogurt?

10. Why was there concern about the hormone BST?

# 15

# The Poultry Industry

## Student Objectives

**When you have finished studying this chapter, you should be able to:**

- Explain the importance of poultry to humans.
- List the different segments of the poultry industry.
- Explain how eggs are hatched commercially.
- Describe how broilers are produced.
- Explain how eggs are produced.
- Discuss the growing of turkeys and other poultry.

## KEY TERMS

game birds
broilers
nutrition
incubator
brooder
layers

Humans have grown poultry for almost as long as they have been involved with agriculture. The earliest archeological evidence reveals that poultry was a part of the human diet. All through the history of the world's great civilizations is evidence that people ate chickens and other birds.

Modern Americans also eat a lot of poultry. Each year, the average person in this country consumes almost 90 pounds of chicken and 22 dozen eggs. The amount of poultry meat consumed

**FIGURE 15-1**
Poultry is lower in calories and cholesterol than beef or pork.
*Courtesy of American Broiler Council.*

is rising each year. The reason is that poultry is considered to be healthier than some other meats. The amount of cholesterol and fat is lower in a pound of chicken or turkey than in a pound of red meat.

The poultry industry includes the raising of chickens, turkeys, ducks, geese, **game birds** (birds hunted for food), and squab (pigeons). By far the largest segment of the industry is the growing of chickens. This industry is divided into two parts—broiler production and egg production.

At one time most farms raised chickens for both meat and eggs. Females were kept to lay eggs and the males were butchered for meat. Today the industry is more specialized. Two different types of birds are produced. Through selective breeding, chickens have been developed that are thickly muscled and produce more meat. Layers have been developed that produce eggs daily. They are more slender than the birds raised for meat. Layers convert more of their energy into egg production than do meat type birds.

**Meat and Egg Producing Chicken Breeds**

**Meat**

**Egg**

**FIGURE 15-2**
Broiler chickens are a different type than layer chickens.

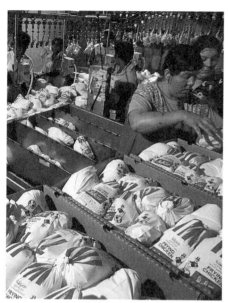

**FIGURE 15-3**
We can grow a heavier broiler in half the time on half the feed than we could in 1950. *Courtesy of James Strawser, The University of Georgia.*

# The Broiler Industry

**Broilers** are chickens that are produced for meat. They are generally marketed at about 6 to 8 weeks of age and weigh 3 to 5 pounds. Since 1950 the production of broilers has increased dramatically. Now we can grow a broiler that is heavier on half the feed in half the time. This is mostly due to research in **nutrition** and management and also to selective breeding. This means a real savings for the consumer. The efficiency with which these animals can be grown means that they can be sold cheaper.

Most of the nation's broilers are grown in the southern states. The leading states in broiler production are Arkansas, Georgia, and Alabama. In these states the winters are mild and the chickens can be grown at a cheaper cost. Most broilers are raised on contract. This means that the producer supplies the house, utilities, and labor. A company supplies the producer with chicks, feed, medication, and other supplies. The company pays the producer for the broilers at the end of the grow-out period. This substantially reduces the risk for the producer.

## Hatching the Chicks

The first step in the production of broilers is the hatching of chicks. Breeder birds are carefully selected for rapid growth, good feed efficiency, and heavy muscling. Almost all of the broilers produced are white. This is because birds that are dark in color have spots of color or pigmentation left where the feathers are removed after slaughter. These spots do not lower the quality of the meat in any way, but consumers are reluctant to buy chicken with spots on the skin.

Eggs are hatched in large commercial hatcheries. Eggs that are to be hatched have to be fertile. This means that the hens have to be bred by roost-

**FIGURE 15-4**
Almost all of the broilers produced are white. *Courtesy of James Strawser, The University of Georgia.*

**FIGURE 15-5**
Hens that lay eggs for hatching are housed in facilities with laying boxes. *Courtesy of James Strawser, The University of Georgia.*

ers or they are artificially inseminated. Hens that lay eggs for hatching are housed in facilities with laying boxes. These boxes are designed to keep the eggs clean. It is essential that hatching eggs be clean. Even a small speck of foreign material on the shell can contain millions of microorganisms that can cause problems. Eggs that are dirty are not used for hatching. Dirt or any foreign material that must be scrubbed from the eggs usually renders the eggs unfit for hatching. The washing or scrubbing of the eggs removes the protective coating from the eggs and presses the dirt into the pores of the eggs. Hatching eggs are not allowed to become wet. Moisture on the eggs can also encourage the growth of bacteria.

Before they leave the farm where they are produced, hatching eggs are sorted to remove dirty, undersized, oversized, misshapen, cracked, or defective eggs. They are then fumigated to kill harmful organisms on the surface of the eggs. This pro-

**FIGURE 15-6**
Hatching eggs must be kept clean and dry. *Courtesy of James Strawser, The University of Georgia.*

**FIGURE 15-7**
Eggs are placed in an incubator and are carefully monitored for temperature and relative himidity. *Courtesy of James Strawser, The University of Georgia.*

cess is precisely regulated to prevent harm to the eggs. The eggs are never allowed to become chilled and are stored at 70 to 80 °F until they are placed in the hatchery.

The eggs are transported to the hatchery on special racks that protect the eggs. At the hatchery the eggs are removed from the carts and placed in the **incubator.** Here the temperature is gradually increased to prevent the eggs from sweating. Sweating is the condensing of water vapor on the surface of the eggs and occurs when the temperature of the eggs is raised too rapidly. If moisture collects on the surface of the egg, bacteria may begin to grow and thrive.

In commercial hatcheries, the eggs are incubated in two separate rooms, the setting room and the hatching room. The eggs are placed in the setting room incubator and closely monitored for temperature and relative humidity. The eggs are turned every day to ensure that a high percentage of the eggs hatch. The eggs remain in the setting room incubator until one to two days before they are ready to hatch. They are then placed in the hatching room. The temperature is lowered slightly and the eggs hatch in chick holding trays.

## Growing the Broilers

After the chicks are hatched they are cleaned and dried. They are then placed in a warm environment. When they are about a day old, they are vaccinated and debeaked. In debeaking, the tip of the beak is removed. This helps to prevent the chicks from pecking each other. It does not harm the chicks and prevents problems throughout the life of the chickens.

The chicks are placed in cardboard or plastic ventilated boxes and shipped to the producers. The chicks are placed in houses that hold from 10,000 to 50,000 or more chickens depending on the size of the house. The houses are well insulated and well lit. Lights are kept on around the clock. This helps prevent a problem called cannibalism. For some reason the chickens may begin to peck certain birds. This causes injury and often death. By debeaking the chicks and leaving the light on, cannibalism can be lowered. The floor is covered with an absorbent litter consisting of shavings, sawdust, or other material.

At the start the baby chicks are placed under a **brooder.** Brooders are large pan-shaped heaters

FIGURE 15-8
Chicks are debeaked to prevent them from pecking each other. *Courtesy of James Strawser, The University of Georgia.*

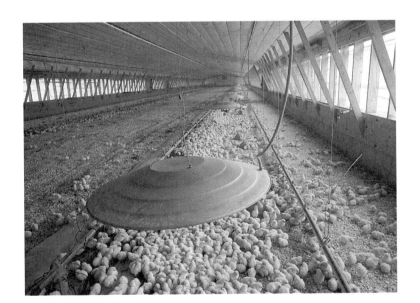

FIGURE 15-9
Chicks are kept under a brooder to keep them warm. *Courtesy of James Strawser, The University of Georgia.*

used to keep the chicks warm during the first days in the house. Brooders are usually suspended from the ceiling and can be raised or lowered depending on the temperature and the size of the chicks.

## Career Development Events

The FFA offers agriculture students opportunities to apply classroom lessons and laboratory skills through competition. Matthew joined the FFA his first year in high school. One of his classes in agriculture was poultry science. Matthew's experience and classroom knowledge prepared him to be a winner in several FFA competitions. Matthew tried out for the Poultry Judging career development event. FFA Poultry Judging is a team event. Matthew made the team along with three other FFA members. The team members spent many hours preparing for the area competition. After school they practiced judging and placing classes of live birds, dressed poultry, interior and exterior of eggs, and processed chicken parts. At home they each studied assigned topics in the poultry science manual that their agriculture teacher would drill them on the following day. Matthew practiced public speaking skills. After weeks of practice and study, the team competed in the area FFA Poultry Judging event. Each of the four members had to judge and give reasons for classes of production hens, pullets, ready-to-cook broilers, chicken patties, interior eggs, and exterior eggs. They also had to identify ten broiler parts and place four cartons of eggs with an explanation of how they placed them. At the end of the event, each FFA member had to answer a 30-question test relating to management practices used in the poultry industry. After all teams complete each section of the event, individual scores were calculated. The top three individual scores for each team were added to determine the team score and placement. Matthew's team placed second and has the opportunity to advance to the state event. Matthew's interest in the processing and marketing of poultry, development of a supervised poultry project, and devotion of time and energy has made him a winner and prepared him for a future career.

The Poultry Proficiency is one of many agricultural proficiency awards presented at the FFA's annual banquet.

Water is supplied by suspended waterers that the chicks quickly learn to use. They peck the nipple on the bottom of the waterer and obtain as much water as they need. Feed is given to the baby

Applicants applying for the poultry production proficiency award must be involved in activities and enterprises requiring the application of classroom lessons to the profitable management of a poultry operation. All FFA members who had a Supervised Agricultural Experience Program (SAEP) in the area of poultry competed for this award. Maintenance of a complete set of records including income and expenses, inventory of poultry and related equipment, and knowledge of the poultry industry helps give students the winning edge.

In the Poultry Judging Career Development Event, FFA members must determine the quality of various poultry products. *Courtesy of Carol Duval, National FFA Organization.*

**FIGURE 15-10**
The chickens are fed through automatic feeders and waterers. *Courtesy of James Strawser, The University of Georgia.*

chicks by hand when they are small but is fed through automatic feeders as the birds get older. The feed is brought to the birds by means of a conveyer chain in the bottom of the trough. Every day the equipment and birds are checked to ensure that the equipment is functioning properly and that the birds are doing well.

The birds are generally kept in the broiler house from seven to eight weeks. At this time they weigh around 4.5 pounds and are ready for market. They are usually caught at night when the birds are less active. They are put into coops and loaded on a truck for transportation to the processing plant. The producer then begins to clean the house to get ready for the next batch of chicks. Since the manure in the litter has a very high concentration of nitrogen and other elements necessary for plant growth, the litter is a valuable source of fertilizer.

## Processing the Broilers

At the processing plant the birds are inspected while they are alive. If they show any signs of ill health they are condemned. This means that they cannot be used for human consumption.

**FIGURE 15-11**
Carcasses are inspected for any signs of disease or other problems. *Courtesy of James Strawser, The University of Georgia.*

After inspection, the chickens are slaughtered and the feathers are removed. The intestines and other internal organs are removed. The carcasses are then washed and inspected. Both the carcass and the internal organs are inspected for signs of disease or any other problems. If any problems are found the carcass is condemned and the meat is not used for human consumption.

After the birds are dressed, they are prepared for sale. They may be packaged whole or they may be cut up into several pieces. The trend is toward selling different parts of the carcass. For example, a package may contain only breasts, legs and thighs, or wings. Some of the broilers may be used for processed meat like chicken franks, sausage, or bologna. Others may end up as chicken pot pies or as prepared frozen dinners.

## The Layer Industry

The other large segment of the poultry industry is the production of eggs. As mentioned earlier, hens or **layers** have been developed through selective breeding that produce many more eggs than their

FIGURE 15-12
Layers are housed in cages where they are fed and watered by automation. *Courtesy of James Strawser, The University of Georgia.*

ancestors. Today the average layer produces over 250 eggs per year. Some layers produce brown eggs and some produce white eggs, but the vast majority of the eggs sold in the United States are white. Consumers simply prefer to buy white eggs. Most of the layers in this country are housed in cages within a large facility. The cages hold from 2 to 12 birds.

The hens are fed by means of an automated conveyer that carries the feed directly in front of the hens. Water is supplied by means of a narrow free-running trough or a nipple waterer that the hens learn to peck to obtain water. As the eggs are laid, they roll onto a conveyer that periodically moves the eggs to a collection point where a worker gathers them and places them in flats. Dirty eggs are separated out and the eggs are refrigerated.

At the processing and packing plant, the eggs are coated with a thin coat of light mineral oil. This coating prevents carbon dioxide from escaping from within the egg. This helps prevent the eggs

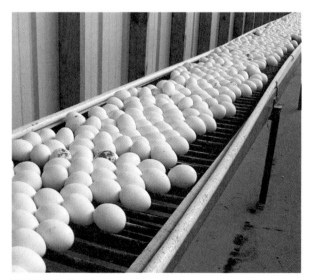

FIGURE 15-13
Eggs are brought from the laying cages on a conveyer. Sometimes they are gathered and packaged at the farm. *Courtesy of James Strawser, The University of Georgia.*

FIGURE 15-14
In a process called candling the eggs are checked for blood spots or cracks. *Courtesy of James Strawser, The University of Georgia.*

from spoiling before they are consumed. The eggs are checked for cracks and interior spots in a process called candling. The eggs are passed over an intense light in a dark room and any blood spots or cracks in the shell will show up in the light. The eggs are then graded according to shape and size. Eggs of similar size are packaged together. The eggs are packaged and sent to the retail market or are sent to a processing plant where they are broken and processed.

## Turkey Production

Turkey production is on the increase in this country. This is due to the many new ways of using turkey meat. Luncheon meat, sausages, wieners, and products resembling ham are made from turkey. Turkey meat is less expensive, and has fewer calories and cholesterol than many other meats. It also tastes good and is nutritious.

Turkeys are the descendants of the wild turkeys that were native to the Americas. Modern turkeys are very different from the wild variety. Like chickens, most domestic turkeys are white. Consumers do not want the coloring in the skin that comes with the bronze-colored turkeys. Also turkeys raised for meat are much more muscular. The modern broad-breasted turkey is much more efficient at producing meat than its wild ancestors.

The majority of the turkeys in the United States are grown in the western part of the North Central region, the South Atlantic region, and the Pacific region. Unlike chickens, turkeys seem better able to tolerate cold weather than hot weather. Most are produced by small operations of 30,000 birds or fewer.

There are two major ways of growing turkeys—in confinement houses and on ranges. Confinement houses offer the advantages of environmental control of temperature and humidity. The open range

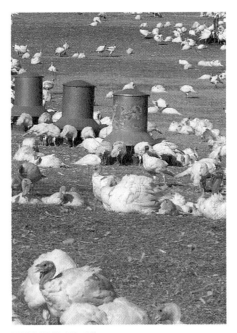

**FIGURE 15-15**
Turkeys may be grown on the range or in a house. *Courtesy of James Strawser, The University of Georgia.*

offers the advantage of being less expensive. Turkeys on range can stay outdoors completely or they are provided with housing where the birds can get shelter when they want. Producers usually move the range every three to four years to help prevent problems with disease and parasites.

## Other Poultry

Although chickens and turkeys make up by far the largest portion of the poultry industry, other types of poultry are also grown. In some parts of the world poultry such as ducks and geese make up a major portion of the total poultry output. In China and Southeast Asia, ducks and geese make up a large part of the overall diet.

In some ways, these birds are actually easier to raise than chickens and turkeys. They withstand harsh weather better and resist disease better than turkeys and chickens. Another advantage of growing geese and ducks is that the feathers are used to make bedding and other goods. In this country most ducks and geese are raised in small flocks by

**FIGURE 15-16**
In some parts of the world, ducks and geese make up a major part of the overall diet. *Courtesy of Robert Newcomb, The University of Georgia.*

**FIGURE 15-17**
Pheasants and quail are raised for use in restaurants. *Courtesy of International Quail, Greensboro, GA.*

hobbyists or part-time producers. Most of the meat sold goes to the restaurant trade or gourmet food market.

About the only other poultry production of any significance is that of growing quail and pheasant. Both of these birds are grown for the restaurant trade and the gourmet food market. Also they are raised for restocking wildlife areas. Each year thousands of quail and pheasants are released in the wild to provide birds for hunting. This helps replenish the areas with game that is difficult to produce in the wild.

## Summary

As the world's population increases, the growing of poultry will become more important. Poultry is efficient to raise and most cultures eat poultry. Many developing countries will depend on poultry to help in the development of their agriculture. In America, we will continue to eat more poultry and poultry products.

# 15 CHAPTER REVIEW

## Student Learning Activities

1.  Make an inventory of all the foods in your house that contains eggs. Do not forget such foods as pastries and mayonnaise. Share your list with the class.

2.  Over a period of a week, keep tract of all the meat products eaten by your family. Calculate the percentage of meals that contain poultry products. Include processed meats such as luncheon meats and hot dogs that may contain chicken.

3.  Research the possibilities of raising exotic birds such as emus and ostriches. Provide a rationale for growing the birds and the advantages and disadvantages. Report to the class.

## True/False

1.  ___ Humans have been growing poultry for almost as long as they have been involved with agriculture.

2.  ___ Each year, the average person in the United States consumes almost 90 pounds of chicken and 22 dozen eggs.

3.  ___ At one time, most farmers raised chickens for both meat and eggs.

4.  ___ Most of the nation's broilers are grown in the northern states.

5.  ___ Almost all of the broilers produced are tan.

6.  ___ The vast majority of the eggs sold in the United States are white eggs.

7.  ___ Turkeys cannot tolerate cold weather as well as chickens.

8.  ___ In China and Southeast Asia, ducks and geese make up a large part of the overall diet.

9. ___ Chickens make up the largest part of the poultry industry. The second most widely produced birds are geese.

10. ___ Each year thousands of quail and pheasants are released into the wild to provide birds for hunting.

# Multiple Choice

1. The meat that is considered to be healthier than most meats is
   a. beef.
   b. pork.
   c. poultry.

2. The largest segment of the poultry industry is raising
   a. turkeys.
   b. fish.
   c. chickens.

3. Raising chickens is divided into two parts, broiler production and
   a. egg production.
   b. candling.
   c. debeaking.

4. Chickens that produce eggs daily are called
   a. layers.
   b. brooders.
   c. hatchers.

5. Chickens produced for meat are called
   a. layers.
   b. hatchers.
   c. broilers.

6. One of the first steps in the management of broilers is
   a. candling.
   b. debeaking.
   c. hatching.

7. Eggs that are to be hatched must be
    a. fertile.
    b. debeaked.
    c. fumigated.

8. At processing and packaging, eggs are lightly coated with
    a. cooking oil.
    b. olive oil.
    c. mineral oil.

9. Candling is a process that checks for defects in
    a. feather pillows.
    b. eggs.
    c. fish.

10. Areas designed to keep eggs clean are
    a. laying boxes.
    b. incubators.
    c. conveyors.

## Discussion

1. How are breeder birds selected?

2. Why are almost all of the broilers white?

3. Why is it essential that hatching eggs be clean?

4. What are the two rooms in a commercial hatchery?

5. Why are lights kept on in a chicken house?

6. Why is the litter in a chicken house a valuable fertilizer?

7. Why are eggs coated with mineral oil?

8. What are the two major ways of growing turkeys?

9. What animals are raised in the poultry industry?

10. What is meant by raising broilers on contact?

# The Science of Aquaculture

## Student Objectives

*When you have finished studying this chapter, you should be able to:*

- Discuss why aquaculture is a growing industry.

- Explain why fish are efficient agricultural animals.

- Explain the steps involved in the production of both catfish and cool-water fish.

- Discuss the production of crustaceans for human consumption.

- Describe the nonfood segments of aquaculture.

One of the newest and most rapidly developing parts of agriculture is that of growing animals in water. This science is called **aquaculture.** Aqua means water and culture means growing or caring for. Many of the foods we eat were grown in the water. You probably ate some type of fish this past week. Perhaps you enjoy seafood such as oysters or shrimp. The amount of fish and seafood eaten by Americans is increasing

FIGURE 16-1
The amount of fish and seafood eaten by Americans is increasing each year. *Courtesy of John Wazniack, Louisiana State University Agricultural Center.*

each year. Experts tell us that eating fish is nutritious and makes a good addition to the diet. It is high in protein and low in calories. The fat content is lower than that of red meat, and this helps prevent health problems. As people become more health conscious, the demand for fish and seafood will increase.

People have grown fish for food for thousands of years. The ancient Chinese raised fish, as did the Egyptians and later the Romans. This provided these people with a ready supply of fish. Those of you who enjoy fishing know that on certain days, even the best angler cannot catch any fish. If the fish have been raised in small ponds, they can be caught at any time.

Only during the past 30 years has aquaculture become a major industry in this country. Up until that time, most of the fish sold in the markets were caught from the wild. Large seagoing trawlers harvested the fish from the oceans. This supplied the world with a relatively inexpensive source of fish. However, as the world population increased and the demand for fish increased, more and more fish were caught from the ocean. A few years ago the maximum amount of fish that could be taken from the

**FIGURE 16-2**
Up until a few years ago, most of
the fish we ate came from the
ocean. *Courtesy of James Strawser,
The University of Georgia.*

oceans in a year was reached. The oceans simply
cannot sustain enough fish to meet the demand.
This created a greater demand for fish than there
was a supply of fish. As a result the price of fish
went up. To provide people with a supply of fish,
fish farms came into production.

At first not much was known about growing fish
for food. But scientists at several of our universities
conducted extensive research in how to produce
fish. This resulted in an industry that is now providing us with an inexpensive supply of fish.

Fish are very efficient farm animals. This means
that a pound of weight can be put on these animals
with less feed than other agricultural animals. Fish
can gain a pound of body weight on less than two
pounds of feed. A steer may require 8 or 9 pounds
of feed to gain a pound of body weight. This is
partly because fish are **ectothermic.** This means
that they depend on their environment for their
body temperature. In other words, the water gives
them their body temperature. Most other agricultural animals, such as chickens, pigs, and cattle, are
endothermic and rely on heat generated from their
feed to maintain their body temperature. Land ani-

**FIGURE 16-3**
Aquatic animals that are suspended in water use less energy moving about. *Courtesy of James Strawser, The University of Georgia.*

mals also have to use a lot of energy moving about. Aquatic animals are suspended in the water and use less energy in moving about. All this saved energy can be put into growing and gaining weight.

## Catfish Production

By far the most important food fish grown in this country is the catfish. In 1960, there were only 600 acres of commercial catfish ponds in the United States. Today there are over 150,000 acres of catfish ponds. These fish are raised in the southern part of the country because they are **warm-water fish.** In fact, catfish grow best when the water temperature is around 85 °F. Mississippi leads the nation in catfish production. For many years the fertile Mississippi delta region produced cotton as the major crop. Today, catfish has replaced cotton as the number one cash producing agricultural enterprise.

Catfish are different from most other fish. They have a smooth skin instead of scales. They also have long whiskers (known as barbells) extending from their mouths. They are meaty and have a mild

**FIGURE 16-4**
Catfish has replaced cotton as the number one crop of the Mississippi delta region. *Courtesy of James Strawser, The University of Georgia.*

**FIGURE 16-5**
Catfish can tolerate relatively low levels of oxygen. *Courtesy of Evelyn Barnes, The University of Georgia.*

flavor. They can be cooked in a variety of ways ranging from deep fat frying to broiling. Also, catfish fit well into farming operations because they are hardy and tolerate relatively low oxygen content in the water.

The production of catfish begins with the breeding process. Through a system called selective breeding, superior fish are used to provide eggs for hatching. Such factors as growth rate and meatiness of the fish are used to determine which fish are superior. Although catfish have been known to reach a weight of over a hundred pounds, the best size for a brood fish seems to be around 3 to 5 pounds.

The fish are placed in ponds less than 6 feet deep called **brood ponds.** These ponds are used to keep fish that breed and lay eggs. Boxes are supplied for the fish to use as a nest for spawning and are submerged. The female lays several thousand eggs (called **roe**) at a time. After the female lays a mass of roe in the nesting box, the male fertilizes the eggs

**FIGURE 16-6**
The female catfish lays a mass of eggs called roe. *Courtesy of James Strawser, The University of Georgia.*

by depositing sperm into the mass of eggs. In the wild the male cares for the nest until the eggs hatch.

In aquaculture operations, the producer takes the eggs from the nest soon after they are laid and fertilized. The eggs are placed in a hatching tank where paddles rotate and gently move the water. After the eggs hatch, the small fish, called **fry,** are placed in long tanks. In these tanks the fry are fed and cared for until they become **fingerlings.** Fingerlings are small fish that are from 2 to 6 inches in length. This size fish is what is used to stock the ponds.

FIGURE 16-7
The life cycle of the catfish.

**Life Cycle
of the Catfish**

**Foodfish**
1 to 2 pounds

**Broodfish**
3 pounds or more

**Stocker**
± 3/4 pound

**Egg**
Hatching occurs
in 7 to 8 days

**Fingerling**
2 to 6 inches

**Sac Fry**

**Advanced Fry**
about 1 inch

The grow-out ponds are about 4 to 6 feet deep and are usually around one to two acres in area. The smaller ponds make harvest easier. In these ponds the fish are fed on a protein-rich diet. The feed floats on the water so the fish have to come to the surface to get it. This allows the producer to see the fish each day as they are fed. An alternative is to raise the fish in cages that are submerged in the water. This gives the advantage of having the fish together in one place. This allows easy feeding, care, and harvest. Also the cage offers protection from predators.

During the period when the fish are growing, the producer keeps careful check on the pond. Fish, like all other animals, must have oxygen. Their oxygen comes from the oxygen that is dissolved in the water. Water can run out of **dissolved oxygen,** and if this happens, the fish will die. The producer checks the oxygen level of the water to ensure that the fish have enough. Loss of oxygen occurs most often during hot, dark nights when plants in the water are not giving off oxygen. The producer can put oxygen back into the water by using mechanical aerators. These devices sling water into the air where oxygen becomes mixed with the water.

**FIGURE 16-8**
Raising fish in cages makes them easier to feed and harvest. *Courtesy of James Strawser, The University of Georgia.*

**FIGURE 16-9**
Mechanical aerators sling water into the air to add oxygen. *Courtesy of James Strawser, The University of Georgia.*

The fish are harvested when they weigh one to two pounds. Consumers prefer this size. The fish are harvested using a large net called a seine. The net is dragged through the water to collect the fish. They

## Career Development Events

**FFA**

The Speciality Animal Production proficiency is one of the many awards available to FFA members. This award is presented to individual FFA members who have effectively produced and marketed such animals as rabbits, fish, dogs, ducks, and bees. The student must have a Supervised Agricultural Experience Project (SAEP) in the selected special animal area. The breeding, feeding, and marketing of fish is an excellent project.

Pam's agriculture teacher invited her to join the FFA when she was in the ninth grade. Pam earned the Greenhand degree and attended FFA camp that first year. Her sophomore year Pam wanted to be a part of the Supervised Agricultural Experience Program (SAEP). After consulting with her parents and agriculture teacher, she decided to grow tropical fish in tanks she kept in her basement. Pam signed up for an aquaculture class and learned how to select, feed, and adjust the environment to produce

healthly stock. She purchased the necessary equipment to perform water quality tests and determine weight gains of her fish. In the second year, Pam produced small fish that were sold to local pet stores. Pam's good management practices produced attractive, healthy fish. With the assistance of her agriculture teacher, Pam continued her aquaculture project for three years. Keeping an inventory of equipment, feed, supplies, and the fish was a constant job. Pam used a computer to track the amount of money she spent and made to calculate each year's net income. At school, Pam won first place in the science fair with a water quality project that she tested in controlled tanks with some of her fish. Pam sang in the school chorus and played soccer for the YMCA. By her senior year Pam had earned her Chapter and State degrees, served as the chapter reporter and participated in several FFA events. Becoming involved in chapter activities, developing an outstanding supervised experience program, keeping accurate records and are then placed in a tanker truck filled with water containing a small amount of salt. The salt has a calming effect on the fish. It is important that the fish reach the processing plant alive. Once fish die,

inventories, showing the practice of required skills, and preparing a neat and complete application are required for winning proficiency awards. Pam won her chapter proficiency award in Specialty Animal Production and received a medal at the annual FFA banquet. Students who win chapter awards can advance to the district and state competitions. The National FFA Foundation awards plaques and prize money for state and national proficiency winners.

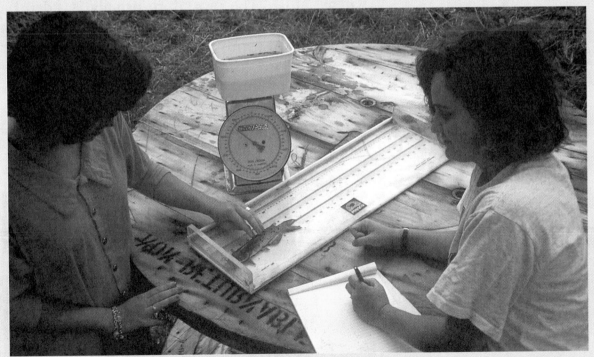

Students with aquaculture projects may qualify for the Specialty Animal Production Proficiency Award. *Courtesy of Blane Marable, Morgan Co. High School, Madison, GA.*

**FIGURE 16-10**
Fish are delivered to the market alive. *Courtesy of James Strawser, The University of Georgia.*

they spoil very rapidly. Packers clean and process the fish as quickly as possible. This ensures that consumers receive a fresh, high-quality product.

## Cool-Water Fish

In the cooler northern portion of our country, trout, Atlantic salmon and striped bass are raised. These are known as **cool-water fish** because they thrive in water that is below 65 °F. These fish require free-flowing clean water. They must also have a higher oxygen content in the water than catfish. For these reasons trout and salmon are usually raised in concrete structures called raceways. Here water flows constantly. This not only helps keep the water clean but helps keep the oxygen levels high.

The spawning of cool-water fish is different than that of catfish. The procedure usually is aided by a technician. When the female is ready to lay eggs the technician gently applies pressure to the lower abdomen of the fish. This causes the eggs to ooze out into a container. A similar process is then applied to the male fish, and sperm is collected. The sperm is then applied to the eggs for fertilization.

**FIGURE 16-11**
Cool-water fish are raised in raceways of moving water. *Courtesy of The University of Georgia.*

The eggs are hatched in jars in which the water is controlled for temperature and oxygen level. When the fry hatch, they are raised in tanks until they are large enough to stock in the raceways.

# Growing Crustaceans

**Crustaceans** are aquatic animals that have an exterior skeleton (exoskeleton) and three body parts—the head, thorax, and abdomen. Included are shrimp, crayfish, and prawns. These animals have been used as food for thousands of years. Although tons of these animals are harvested in the wild each year, lots are raised by producers.

Crayfish are raised in several states. The leading states are Louisiana and Texas. Most are raised in rice fields. The rice fields have to be flooded for a length of time, and during this time the crayfish are raised in the water. They eat the decaying plant material left from growing the rice. They also eat worms and insect larvae. Crayfish are harvested using baited traps that are placed in the water. They are then placed in wet bags that are kept moist and cool and sent to the processing plant. Many of the crayfish are sold live.

Worldwide there are over 35,000 shrimp farms. The major shrimp producing countries are located in the eastern hemisphere. China, Thailand, the Philippines, Taiwan, and Japan are major producers. Only about 20 percent of the world's supply is produced in the western hemisphere. Central and South America are the leading regions.

Shrimp farms are generally located near the ocean because shrimp have to be raised in salt water. The producers usually make use of the rising tide or flood gates to fill impoundments with salt water. Many of the same technologies used with fish, such as mechanical aeration, are used in shrimp production. The shrimp are harvested by using

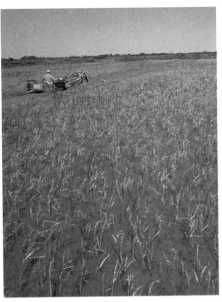

**FIGURE 16-12**
Crayfish are raised in flooded rice fields. *Courtesy of John Wazniack, Louisiana State University Agricultural Center.*

**FIGURE 16-13**
Shrimp are raised in aquaculture operations also. *Courtesy of John Wazniack, Louisiana State University Agricultural Center.*

seines. They are then put on ice and shipped throughout the world.

Prawns are sometime referred to as freshwater shrimp. These animals have to be grown in very warm water (75 to 95 °F). The young (called larvae) have to be grown in brackish water. Brackish means that the water has a salt content lower than ocean water and higher than fresh water. At the proper stage they are placed in fresh water. Here they are raised until harvest.

## Ornamental Fish

One of the fastest growing segments of aquaculture is the growing of **ornamental fish** or fancy fish. A favorite hobby of many people is raising tropical fish in home aquariums. These fish come in a wide variety of sizes, shapes, and colors. These fish can be very valuable. Because of their value, extra care and expense can be invested in their production. Most tropical fish are raised in what is called a closed-loop system. This system consists of a large tank, a filtration system, and an aeration system. The fish must be kept at a constant temperature and the wa-

**FIGURE 16-14**
The raising of ornamental fish is a growing industry. *Courtesy of James Strawser, The University of Georgia.*

ter must be clean. The spawning process is carefully monitored because many tropical fish eat their young. Once the young are hatched they are grown to a certain size and shipped to pet stores all around the world.

An ornamental fish called the Koi is gaining in popularity. This fish is a species of carp and has varying patterns of gold, white, and black colors. They are used in landscape ponds to decorate yards. These fish can be quite expensive. Breeding fish of the right color and color pattern have been known to sell for millions of dollars.

## Summary

Aquaculture is a departure from the traditional concept of agriculture. This relatively new industry will likely grow even more as we enter the 21st century. Some day aquaculture crops such as lobsters, bullfrogs, and other animals now harvested from the wild may be widely produced. Agricultural research will find a way to make this happen.

# 16 CHAPTER REVIEW

## Student Learning Activities

1. Visit the local pet shop and interview the manager. Determine where the shop gets the fish it sells. Ask how they are ordered, shipped, and received. Also, find out how the fish are cared for. Report your findings to the class.

2. Visit the grocery store and compare the price of farm raised fish with fish caught from the wild. Also compare the cost to beef, pork, and poultry.

3. Go to the library and research a particular type of fish. Determine if the fish might have potential for commercial production. Write your findings in a report.

# True/False

1. ___ One of the newest and most rapidly developing parts of agriculture is that of growing animals in water.

2. ___ Fish are not very efficient farm animals.

3. ___ Aquatic animals use less energy than land animals.

4. ___ The most important food fish grown in this country is the catfish.

5. ___ Catfish has replaced cotton as the number one cash crop in the Mississippi delta.

6. ___ Cool-water fish require a low oxygen content in water.

7. ___ Shrimp must be raised in salt water.

8. ___ One of the fastest growing segments of aquaculture is the growing of ornamental fish.

9. ___ Fish is high in protein and low in calories.

10. ___ The fat content of fish is lower than that of red meat.

# Multiple Choice

1. The state that leads the nation in catfish production is
    a. Missouri.
    b. Mississippi.
    c. Florida.

2. The production of catfish begins with this process:
   a. breeding.
   b. stocking.
   c. spawning.

3. Small fish that are from 2 to 6 inches in length are called
   a. crustaceans.
   b. fingerlings.
   c. prawns.

4. Most tropical fish are raised in what is called a
   a. closed-loop system.
   b. brood pond.
   c. raceway.

5. Most crayfish are raised in _____ fields.
   a. corn
   b. wheat
   c. rice

6. Catfish eggs are called
   a. roe.
   b. seine.
   c. prawns.

7. Cool-water fish thrive in water that is below
   a. 45°F.
   b. 65°C.
   c. 65°F.

8. Trout and salmon are usually raised in concrete structures called
   a. prawns.
   b. raceways.
   c. closed-loop systems.

9. Crustaceans are aquatic animals that have
   a. endoskeletons.
   b. epidermis.
   c. exoskeletons.

10. The best size for a brood fish is around
    a. 3 to 5 pounds.
    b. 7 to 8 pounds.
    c. 4 to 6 pounds.

# Discussion

1. Why is aquaculture such a growing industry?

2. Why did fish farms come into existence?

3. Why are fish considered efficient farm animals?

4. Why are catfish so popular?

5. What are raceways?

6. What are crustaceans?

7. What are the major shrimp producing countries?

8. What is brackish water?

9. When does the loss of oxygen occur most?

10. How can oxygen be put back into water?

# Companion Animals*

**KEY TERMS**

companion animals
service animals
hippotherapy
zoonoses
parasite

Have you ever considered **companion animals** to be a part of agriculture? We would all agree that people who own pets are involved in raising the animals. Companion animals play an important role in the agricultural industry. People have always had a relationship with animals. In the last 50 years, our relationship with many domestic animals has changed. As our society has become more urban, the importance of these animals is measured less by the work

*This chapter was contributed by Jean Kilnoski.

they perform for us. Animals still have an important place in the lives of many people, however. They are more than pets; they are friends and companions. It is this special relationship between the animal and its owner that identifies a companion animal.

Companion animals are an important part of our society. In fact, pet ownership is at an all time high. Almost 60 percent of American households have at least one companion animal. Virtually any kind of pet can be considered a companion animal. Dogs and cats are the most common, but companion animals may also be horses, rabbits, hamsters, gerbils, guinea pigs, and ferrets. The companionship bond may also be developed with some unusual pets, such as pot-bellied pigs, reptiles, or fish. This wide range of pets provides companionship for people of all ages.

In recent years, cats have outnumbered dogs as the favorite pet. There are about 60 million cats and 52 million dogs in American households. Many cat owners have more than one cat, so although they are more numerous, dogs are found in more households.

The popularity of dogs and cats as companions is due to a combination of factors. One of these is

**FIGURE 17-1**
Dogs and cats are the most popular companion animals, but animals such as ferrets are also popular.

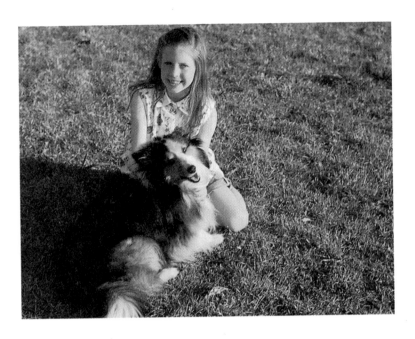

FIGURE 17-2
Dogs are able to express themselves to their owners.

the ability of dogs and cats to form lasting social bonds with humans. They are able to establish a close relationship with their owners. They can also be house-trained fairly easily. House-training is impossible with many companion animals. Dogs and cats also have the advantage of being large enough for humans to interact with and play with and yet small enough to be kept in the house.

The most important reason for the popularity of dogs and cats is their ability to express themselves to their owners. Both dogs and cats have a number of gestures and sounds they use to communicate with their owners. People tend to attribute human feelings and emotions to their pets because of this ability to be expressive.

The fastest growing category of pets in the United States is reptiles. There are more than seven million pet reptiles living in homes in the United States. Iguanas are the most popular pet among reptiles. These lizards are cat-sized when full grown. They are clean, odorless, and can be house-trained. This category also includes snakes and tortoises, as well as amphibians such as frogs and salamanders.

FIGURE 17-3
Reptiles are rapidly becoming some of the more popular pets.

# Health Benefits

Pets are more than just companions—they are good for people's health. Scientists are now discovering that living with a pet contributes to the physical and mental well-being of humans. Relationships with animal companions appear to be beneficial to humans because they are uncomplicated. Animals are accepting and attentive, and respond to affection. They are not judgmental, they never talk back, they never criticize or give orders. Pets give people something to be responsible for and make them feel special and needed.

## Career Development Events

**FFA**

The FFA offers many opportunities for students to develop and display projects. Students even have the opportunity to develop their own programs and the area of companion animal production is well suited for this purpose. If you are involved with the training, grooming, care, or production of companion animals, you might want to organize an event that will showcase your animals.

The first step would be to locate others in your chapter with interest in companion animals. With the help of your advisor, form a committee to discuss the event. Begin by developing objectives and determining what you want to accomplish with the event. The objective might be to have a demonstration of grooming techniques, a competition in grooming, or a competition in the training of companion animals. Use your imagination—there are lots of types of events you could organize.

Once you determine what you want to accomplish, organize the procedures and establish guidelines to govern the event. Get everyone in your chapter involved in the planning and activities and with your advisor's help, organize the members into committees to get every-

Companion animals are a good influence on children of all ages. They help children develop a sense of security. They have been used to encourage shy or withdrawn children to open up. Children who are normally hyperactive often become calmer around a companion animal.

New evidence of the beneficial effects that companion animals have on human health is continually being discovered. We now understand that pets make people feel good, and a sick person who feels good mentally is likely to get better faster. People who have pets report fewer minor health problems, like colds and flu. Studies have also shown that

**FIGURE 17-4**
Pets can make people feel needed.

thing done. You might even get the local FFA Alumni Affiliate involved, which offers support for the FFA programs.

As your event becomes successful, you may wish to invite other chapters to participate. This was the way most of the major FFA events began. The FFA is growing and looking for new events and you just might have an idea that would grow into a national event.

Students who raise companion animals as an SAEP may enter the Specialty Animal Production Proficiency Award.

**FIGURE 17-5**
Watching fish in an aquarium can help lower a person's blood pressure.

petting a dog or a cat or watching fish in an aquarium can help lower a person's blood pressure and heart rate.

Due to the increasing mobility of today's society, many elderly people no longer have family members living close by. Therefore, companion animals play an important role in the lives of a number of older individuals. Many of them would feel isolated and alone without their pets. Pets provide them with companionship and a sense of being valued and needed. Older people may actually live longer, healthier lives because of their relationship with their companion animals.

Today about 50 percent of nursing homes use animals in some capacity. Bird aviaries and aquariums are popular among residents, as well as cats, rabbits, and guinea pigs. It is not recommended that dogs live in nursing homes full-time because they tend to be overfed by the residents, who cannot seem to resist feeding them cookies and treats. Today many volunteers take their dogs to visit the residents of nursing homes and hospitals in their community.

# Service Animals

Some companion animals serve the dual roles of companionship and practical assistance. They are known as **service animals.** A number of agencies in the United States train assistance dogs. These dogs are trained to give more independence and mobility to people with disabilities. Dogs serve as the eyes, ears, or legs for thousands of people. Training usually takes between four and eight months, depending on the difficulty of the tasks that must be learned and the aptitude of the dog. Although training an assistance dog can cost thousands of dollars, many agencies provide them to people who need them at little or no cost.

The best known example of an assistance dog is the guide dog for the blind. The most commonly used breeds are German shepherds, Labradors, and retrievers. To work as a guide dog, the dog has to be exceptional. It has to be able to walk through crowds, climb stairs, ride on elevators, ride on buses and in cars. Most importantly, the dog has to be able to think for itself. It must learn to disobey a command if it could bring harm to its master.

FIGURE 17-6
The German shepherd is a popular breed used as guide dogs.

Many of the organizations that train guide dogs have volunteers raise the puppies for the first year. Some of the volunteer puppy raisers belong to the 4-H organization. The volunteers are not expected to train the puppies as guides, but they are required to follow some basic rules. The puppies must be exposed to many of the activities that they will need to handle with ease as a guide. The puppy raisers are encouraged to take the puppies to the mall, to the park, to nursing homes, schools, or any other setting that involves the public. The puppies must be kept on a lead in public, and they must sleep next to the bed of the volunteer, just as it will when acting as a guide.

The volunteers are warned not to play ball, tug of war, or other games with the puppies that could turn into bad habits when they become guide dogs. The puppy raisers must never feed their pets human food. Guide dogs must not be tempted by the smell or presence of human food since they may accompany their owners to restaurants. They cannot be jumping up on tables begging for food.

Only about half of the puppies raised to be guide dogs successfully complete the training. During training the dogs learn to work in a harness, and they learn commands like "Forward" and "Find the door." The dogs are trained through repetition and praise. They learn to ignore crowds, noises, squirrels, and cats. The dog must learn that it is as wide and as high as a person. It must be aware of obstructions that it can walk around or under, like awnings or branches that would impede its owner. A guide dog wearing a harness is on duty and should never by petted by other people. When the dog is out of the harness, the dog is like any other family pet.

Hearing-ear dogs are trained to listen for people who cannot. Also called signal dogs, these animals can respond to more than 30 common household

**FIGURE 17-7**
Only about half of the puppies raised to be guide dogs successfully complete the training. *Courtesy Guide Dog Foundation for the Blind, Inc.*

sounds like doorbells, telephones, alarm clocks, and fire alarms. They can even be trained to respond to a crying baby. The dogs alert their deaf or hard-of-hearing owners by walking back and forth from the source of the noise to the owner. Signal dogs are taught not to bark when alerting their master to a sound, since they would be unable to hear the barking. The dogs can also be trained to respond to sign language commands. Because the size of the dog is not important, hearing-ear dogs are usually mixed breeds and are often rescued from local animal shelters.

Service dogs are trained to help people who use wheelchairs or have spinal injuries. They are able to respond to more than 40 different commands. Service dogs can open doors, work light switches, pull emergency cords, and pull wheelchairs. Each dog may be trained a little differently in order to address the needs of the individual who will own the dog. Service dogs need to be large and are often retriever breeds.

Another companion animal that has been beneficial to humans is the horse. Horses are used in programs of physical therapy for people with disabilities. This type of therapy, called **hippotherapy,** can be used with people of all ages, but it is especially helpful for physically challenged children. The gait of a horse simulates the motion of humans as we walk. When we walk, our bodies move from side to side and up and down. Riding a horse recreates that sensation in people who are unable to walk unassisted.

Through hippotherapy, physically challenged individuals improve their balance, posture, strength, coordination, and muscle flexibility. In addition to the physical benefits, the riders also gain confidence. Hippotherapy can provide them with a whole new perspective and a sense of freedom. A horse can take its rider where no wheelchair could go. Volunteers

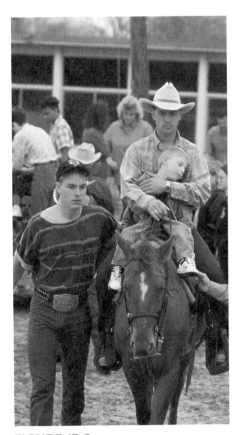

**FIGURE 17-8**
Horses help people who are unable to walk unassisted. The gait of the horse resembles the motion humans make when they walk. *Courtesy of Cliff Ricketts.*

are used in the care and maintenance of the horses as well as for lessons. Hippotherapy programs offer many opportunities to local agriculture programs and youth organizations.

## Diseases and Afflictions

Unfortunately, pets can give us more than companionship. Each year pets pass along infectious diseases to thousands of Americans. There are approximately 30 varieties of pet-borne illnesses, which are called **zoonoses.** Zoonoses are diseases and infections that can be transmitted from animals to humans. They can be passed on through direct contact with the animals or acquired indirectly through contact with animal feces or other contaminants. However, most diseases passed from animals to humans are easy to avoid and are treatable. Good hygiene and safe handling procedures should always be practiced when working with animals. Cats and dogs are responsible for the majority of zoonoses, but birds, fish, and turtles are also culprits.

Rabies is the best known and the most feared example of a zoonosis. It is contracted through the

**FIGURE 17-9**
Good hygiene and safe handling practices should be practiced when working with animals. *Courtesy Warren,* Small Animal Care & Management, *1995, Delmar Publishers.*

**FIGURE 17-10**
Dog owners should make sure their pets are regularly wormed.

saliva of rabid animals. Although this disease is rare in pets, rabies is increasing in the wild animal population. Because pets may come into contact with wild animals, they should be vaccinated against the disease.

The common round worm of dogs is a **parasite** that can infect humans. A parasite is an organism that lives off of or at the expense of another organism. The parasite is transmitted through contact with the animal's feces or with contaminated soil. Children playing in areas frequented by dogs are especially at risk. Dog owners should make sure their pets are regularly dewormed.

Toxoplasmosis is a parasitic disease that can be transmitted by contact with cat feces. Toxoplasmosis can be especially dangerous to pregnant women, so most veterinarians recommend pregnant women not clean cat litter boxes.

Psittacosis, or parrot fever, is a disease transmitted by parrots, budgerigars, and other related caged birds. Humans can be infected by contact with the feces of contaminated birds. In handling birds and cages, dust masks or protective face shields should be used.

Ringworm is not a worm but a fungus that results in a skin aggravation in humans. It is primarily passed to humans by kittens and puppies. The animal appears unaffected because the fungus infects only the animal's fur. It is passed to you when you handle your pet. It is more common in children because adults seem to become more resistant to it with age.

Rocky Mountain spotted fever and Lyme disease are tick-transmitted diseases that can affect both humans and animals. Ticks are found in grassy, wooded areas, or they can be brought into the household by dogs and cats that have been outside. Both diseases are treatable with antibiotics. Normal grooming of animals after they have been outside will help locate and eliminate ticks.

Infections due to animal bites and scratches are another concern. The potential for infection varies. Less than 5 percent of dog bites become infected, but up to 50 percent of cat bites do. Cat scratch fever is associated with cat scratches or bites. This disease is not serious and can be treated with antibiotics. Safe handling techniques are an important

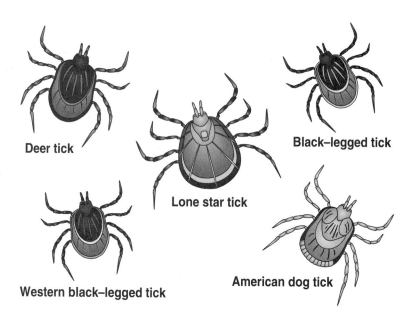

**FIGURE 17-11**
Ticks can transmit Lyme disease.

Deer tick

Lone star tick

Black–legged tick

Western black–legged tick

American dog tick

measure to prevent bites and scratches, as well as injury to the animal. Prompt and thorough washing of pet bites and scratches with soap and water is always important.

Allergies are probably the most common afflictions that occur when humans have contact with animals. Many people develop allergies to animal hair and their dander, or flaking skin. Allergies cause hay fever-like symptoms that can occur in both children and adults.

## Industry

One measure of the value people place on pets is the amount of money they are willing to spend on them. Americans spend over $20 billion a year on companion animals! Dog and cat food purchases account for about $8 billion of the total spent on pets. By comparison, people spend approximately $1 billion on baby food. The companion animal industry contributes to the economy through the sale of pets, pet food, accessories, and veterinary care.

The pet food industry utilizes many of the by-products and surpluses of the human food industry. The main ingredient in dry dog foods is grain, soybean meal, or wheat millings. The main ingredients in canned pet foods are meat by-products, which may include the waste products of meat that was processed for human consumption. The composition of pet food is carefully formulated to meet the nutritional needs of the animals. Canned, semi-moist, and dry foods are equally nutritious, but the canned varieties generally contain a higher percentage of protein and fat. Consequently pets should be fed smaller amounts of canned food or mix dry foods with canned.

Cats have special dietary needs. Pet foods formulated for cats have a much higher protein and fat content than any type of dog food. Dogs often

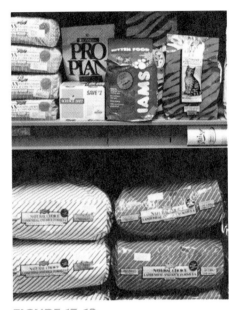

**FIGURE 17-12**
The pet food industry produces a wide variety of pet foods. *Courtesy Warren,* Small Animal Care & Management, *1995, Delmar Publishers.*

**FIGURE 17-13**
Cats require a diet higher in protein and fat than dogs. *Courtesy of James Strawser, The University of Georgia.*

seem to prefer cat food because of the high fat content. For healthy, active dogs, nibbling a little cat food is not a problem, but they will become obese if allowed to eat it regularly. In contrast, cats should not be allowed to eat dog food because it does not contain all of the nutrients that a cat needs to stay healthy.

The supply of pet accessories is also an important contributor to the economy. A companion animal may need any number of essential items such as a feeding bowl, dish, or trough; a cage, hutch, or kennel; and chew bones, scratch poles, and toys. Many pet owners also indulge in luxury items such as heated pet beds, jewelry, pet clothing, and car seats.

## Health Care

Americans spend over $10 billion every year on the health care of their companion animals. There are approximately 40,000 veterinarians in the United States; one-third of them treat small animals exclusively. Like all veterinarians, those who specialize in the care of companion animals perform a wide variety of tasks every day. They treat animal injuries, set broken bones, immunize healthy animals against disease, and perform surgery.

Veterinarians today can successfully perform hip replacements and kidney transplants on companion animals. They have also performed balloon angioplasty to open clogged arteries, open-heart surgery, and dental surgery. In recent years, medical care for animals has become as highly technical as medical care for humans. In fact, new medical procedures are perfected on animals, so in some instances animals may get even more advanced care than humans. However, the costs of these health care advances may be more than some pet owners are willing or able to bear.

**FIGURE 17-14**
One third of all veterinarians in the United States treat small animals exclusively. *Courtesy Warren, Small Animal Care & Management, 1995, Delmar Publishers.*

Veterinarians stress preventive measures when they counsel pet owners. They encourage the pet owner to include vaccination programs and regular dental exams in their pet's health care plan. According to veterinarians, one of the most common problems they see in dogs and cats today is obesity. Half of American dogs and almost one-fourth of cats are obese. The animals suffer from overfeeding and a lack of exercise. Some animals simply do not get enough attention. When animals get bored they have a tendency to eat too much, just like people. A sound diet and daily exercise routine should also be a part of the pet's overall health care plan.

## Responsible Ownership

Responsible ownership of a companion animal requires a long-term commitment and a realistic assessment of the needs of the pet. A number of factors should be considered by the prospective pet owner that will contribute to the well-being of the pet. The first is the amount and kind of space in which the pet will live. A lively hunting dog does not belong in an apartment. Likewise, an easily frightened pet should not be kept in a household

FIGURE 17-15
A lively hunting dog does not belong in an apartment.

with several lively, noisy children. No dog should be tied to a tree and left alone all day, every day. Animals that are forced to live in environments that do not suit their natures may exhibit behavioral problems.

Although many households want to have dogs, busy families that do not have the time or space to devote to a dog have many other options to consider. Rabbits, hamsters, and guinea pigs can be affectionate, cuddly pets and do not require the time or the space of a dog. Reptiles, such as geckos, monitors, and iguanas, are not soft and cuddly, but many of them do seem to enjoy contact with people, probably because of human body heat. Reptiles do not require much space and may even be kept in apartments that do not allow cats and dogs.

Responsible owners of companion animals also need to consider the problem of pet overpopulation. Millions of unwanted and homeless animals are born each year. This overpopulation creates problems for society and leads to suffering for many companion animals. To solve this problem, the number of puppies and kittens born must be reduced.

**FIGURE 17-16**
Small animals such as mice do not require much space.

At least ten million dogs and cats are destroyed, or euthanised, each year by animal pounds and shelters. Future pet owners are encouraged to consider rescuing an animal by adopting a pet from their local animal shelter. It is the responsibility of current pet owners to help control this problem by neutering and spaying their pets.

Spaying and neutering are surgical procedures that prevent animals from reproducing. Spaying is performed on females and neutering is performed on males. The animals are anesthetized so they do not feel any pain during the surgery. They may experience some discomfort for a day or two after the procedure.

Spaying and neutering a pet can result in a healthier, happier animal. Spayed females are less susceptible to some infections and cancerous growths. Neutered males are less likely to display aggressive behaviors, or to roam.

## Summary

Companion animals represent a large and growing segment of animal production. The many segments of this vast industry have a positive effect on the economy. It is different from most of the agricultural industry in that it is not involved with the production of food and fiber. It does, however, provide a valuable product for people who enjoy animals as companions.

# 17 CHAPTER REVIEW

## Student Learning Activities

1.  Visit the local pet store and interview the owner as to where the animals come from. Research how the animals are produced.

2.  Make a list of all the money spent on your pet during the last year. Determine how many different industries were supported by the expenditures.

## True/False

1.  ___ There are more cats than dogs in the United States.

2.  ___ Dog food can also be fed to cats with no ill effects.

3.  ___ A companion animal may be a snake.

4.  ___ A zoonosis is a disease that is passed from a caged animal to a companion animal.

5.  ___ Lyme disease is contracted through contact with cat feces.

6.  ___ Rabies is relatively uncommon in household pets.

7.  ___ Veterinary technology tends to lag far behind human medical technology.

8.  ___ Most dog bites become infected.

9.  ___ The fastest growing category of pets in the United States is reptiles.

10. ___ Spaying is a surgical procedure performed on female animals.

# Multiple Choice

1. Ringworm is caused by
   a. parasites.
   b. fungus.
   c. worms.

2. Signal dogs are trained to assist
   a. people in wheelchairs.
   b. the blind.
   c. the deaf.

3. The main ingredient in dry dog food is
   a. grain.
   b. chicken.
   c. meat by-products.

4. Hippotherapy is particularly beneficial for
   a. the elderly.
   b. physically challenged children.
   c. the deaf and hard-of-hearing.

5. Volunteers who raise puppies for guide dog programs should try to expose the puppy to
   a. games like tug-of-war.
   b. human food.
   c. crowds of people.

6. Neutering is a surgical procedure that is performed on
   a. male animals.
   b. reptiles.
   c. female animals.

7. Service dogs are trained to help
   a. the blind.
   b. the sick.
   c. people in wheelchairs.

8. The best known and most feared zoonosis is
   a. rabies.
   b. Lyme disease.
   c. ringworm.

9. It is not recommended that dogs live in nursing homes full-time because
   a. they tend to be aggressive toward elderly people.
   b. they tend to be overfed by the residents.
   c. companion animals are not beneficial for elderly people.

10. Psittacosis is transmitted by
    a. contact with the feces of some caged birds.
    b. ticks.
    c. parasites.

# Discussion

1. Why is spaying or neutering a pet an important part of responsible pet ownership?

2. How does the companion animal industry contribute to the economy?

3. What are the differences between dry and canned pet foods?

4. Why are dogs and cats the most popular companion animals?

5. What is a companion animal?

6. How do companion animals affect the health of humans?

7. Why are companion animals beneficial to elderly people?

8. What are zoonoses?

9. How do assistance dogs help people?

10. What is hippotherapy?

# 18

# The Ethical Treatment of Animals

## Student Objectives

*When you have finished studying this chapter, you should be able to:*

- Distinguish between arguments of animal rights and animal welfare proponents.

- Explain why producers treat their animals well.

- Discuss why people may feel that agricultural animals are mistreated.

- Give reasons why confinement operations are not cruel to animals.

- Discuss why the use of management practices are humane.

- Discuss the use of animals for experimentation.

People have always used animals. Since humans first appeared on the earth they have hunted and killed animals for food, shelter, and clothing. Very early they began to raise animals in addition to hunting them. From the

**FIGURE 18-1**
Today relatively few people really understand how agricultural animals are raised. *Courtesy of James Strawser, The University of Georgia.*

earliest recorded history to today people have controlled the lives of animals they raised.

In years past most people were familiar with farm animals and understood how they were raised and sold for slaughter. Today in the United States only a small percentage of the people really understand about the raising of farm animals. This creates an atmosphere where much of what happens can be misunderstood. Some people see the raising and management of animals to be cruel. They feel it is unjust to grow and slaughter animals for our use.

There are at least two lines of thinking concerning the well-being of animals. The first is generally referred to as those who endorse **animal welfare.** These people feel that it is all right to raise animals for human use. They say that animals should be cared for properly and that the animals should be comfortable in their surroundings. The other group, referred to as **animal rights** advocates, feel that animals should be free and that they have as much right to live as do humans. They uphold the notion that all animals that are kept in pens, cages, or pastures should be set free.

# Animal Welfare

Producers of agricultural animals would be the first to agree that animals should be treated well. The problem comes about when animal welfare activists disagree with producers over the question of what is proper treatment. Producers know that animals have to be cared for properly. Animals that are abused or mistreated in any way will not grow very well. Producers raise animals for profit. Animals that are miserable in their surroundings do not eat, grow, and reproduce efficiently. Unless animals grow fast and reproduce, the producers cannot make much profit from the animals. Because of this it is in the best interest of the producer to treat the animals well. Besides this fact, most producers raise animals for a living because they like animals. For thousands of years humans have almost literally lived with their animals to care for and protect them. Sheep herders have always had a reputation for caring for their flocks, often at the risk of their lives.

**FIGURE 18-2**
Producers know that to make a profit animals have to be well cared for. *Courtesy of James Strawser, The University of Georgia.*

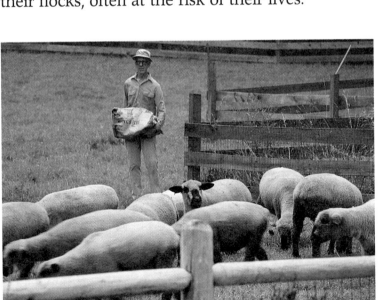

**FIGURE 18-3**
Sheepherders have always had a reputation for providing complete care for their flocks. *Courtesy of James Strawser, The University of Georgia.*

Millions of dollars and countless hours of time have gone into research to determine what makes animals comfortable. Modern facilities have been designed to make animals as comfortable as possible. Housing is built so that the space for the animals is not too warm or too cold. Producers realize that if animals are too hot or too cold, they are stressed. Stress causes animals to use energy that could otherwise be used for putting on weight.

## Confining Animals

Animal welfare proponents sometimes object to putting animals into **confinement operations.** They say it is wrong to give animals only a limited space to live. They point out examples of pigs being raised their entire lives in small pens, laying hens being confined to cages, and sows being kept in farrowing crates. The animal welfare activists contend that animals are being mass produced in what they call corporate-owned **farm factories.**

Producers point out that most of the animals raised are produced on family farms and not on corporate-owned farms. They also argue that facilities are designed specifically for certain animals. For

FIGURE 18-4
Some people feel it is cruel to raise animals in a confinement operation. *Courtesy of James Strawser, The University of Georgia.*

example, a hog house is designed for hogs. This design takes into account the amount of space that a hog needs to be comfortable. It also accounts for the nature of the animal. Pigs are sociable animals and like to be around other pigs. To separate them into isolated spaces where they are by themselves would stress the animals. Research has shown that most animals do better when raised with other animals of their own kind.

The use of **farrowing crates** is another example of a facility designed and used for the betterment of the animal. To those who do not understand the purpose and use of farrowing crates, it might look cruel. Sows are placed in the crates about a week before they give birth. They remain in the crate for about a week after the pigs are born. The sow cannot turn around in the crate and has only room to lie down. The purpose is to prevent her from crushing her piglets as she lies down. The crate is designed so that the sow has to lie straight down instead of flopping onto her side. This slower movement allows the little pigs time to move before the mother lies on them. The sows are given plenty of feed and water and the area is kept dry and

**FIGURE 18-5**
Sows are kept in farrowing crates to protect the baby pigs. *Courtesy of James Strawser, The University of Georgia.*

warm. The sows seem to be in no discomfort and many young pigs are prevented from being crushed to death.

Animals have been selectively bred to live in confinement. This means that they are quite different from their ancestors that roamed freely. Layers that are kept in cages would probably not do well roaming free. The cages are designed for cleanliness and comfort for the hens. They have an abundant source of feed and water and the temperature in the laying houses is carefully controlled.

## Career Development Events

**FFA**

You already know that agriculture is more than farming. Any students interested in a career in livestock should strengthen their knowledge and experience by becoming active in the FFA. The FFA develops competent and aggressive leaders who learn how to speak out and debate issues such as animal welfare. Agriculture education and the FFA provide hands-on experience in practicing skills of health management with animals. This experience gives you a better understanding of animal welfare issues.

Animal welfare has become an issue of debate in local, state, and national agricultural meetings around the world.

To speak out and present your views requires the use of correct parliamentary procedure. All FFA members should learn how to conduct and participate in local meetings. Parliamentary procedure teaches the student an orderly method of conducting business that assures that all sides of an issue are treated fairly. Most FFA chapters conduct a local Parliamentary Procedure Event where six students make up a team. The team is given tasks to perform using correct parliamentary procedure. In preparing for the event, team members must learn proper use of the gavel, voting procedures, order of motions, and other requirements for properly conducting a meeting. Team members must individually answer questions covering basic par-

Animals that are kept in confinement can receive better care than animals that roam free. Parasites have always been a problem to animals. They cause stress and discomfort and may even shorten the life of the animal. When animals are kept in confinement, producers make certain that the animals are kept free of both internal and external parasites. In addition, the animals do not have to worry about predators like animals in the wild or those running free. The producers see the animals every day and care for their needs.

**FIGURE 18-6**
Hens that are kept in cages have been bred for this type of life. *Courtesy of James Strawser, The University of Georgia.*

liamentary law and perform as a team. Parliamentary procedure is an effective method of addressing issues while respecting the vote of the majority and the rights of the minority. The chapter winning team will have the opportunity to compete in area, state, and national FFA Parliamentary Procedure events.

FFA Activities such as public speaking help prepare students to address issues such as animal welfare. *Courtesy of Carol Duval, National FFA Organization.*

## Management Practices

Another problem perceived by animal welfare activists is that of the management practices used by producers. They say that practices like branding, castration, docking, and dehorning are cruel because they put the animal through a lot of pain.

Producers point out that these practices are done for the benefit of the animal. For example, the removal of horns from an animal may cause temporary pain during and after the operation. However, the purpose of horns is for self-defense. Animals under the care and protection of humans have no need for horns. During mating or other times the animals may use the horns on each other. If they are removed at an early age, there is far less likelihood they will injure each other.

Castration will also prevent the animals from fighting. Neutering makes animals more docile and

**FIGURE 18-7**
Management operations are often viewed as being cruel. *Courtesy of Cooperative Extension Service, The University of Georgia.*

**FIGURE 18-8**
Domestic animals have no need for horns. *Courtesy of James Strawser, The University of Georgia.*

less aggressive. Normally, male animals fight to establish mating dominance. Castration prevents this behavior and prevents injury both to animals and humans. Practices like docking help keep the animals clean. Remember that all the practices used by producers are intended to help and not harm the animal. It is in the best interests of the producer and the animal to keep the animal healthy, comfortable, and contented.

Some practices, like hot-iron branding, are used less now than in the past. Newer techniques, such as freeze branding and tattoos, are being used. These marking techniques cause less stress to the animals.

Several years ago, the National Cattlemen's Association adopted the following statement of principles on animal care, environmental stewardship, and food safety:

*I believe in the humane treatment of farm animals and in continued stewardship of all natural resources.*

*I believe my cattle will be healthier and more productive when good husbandry practices are used.*

*I believe that my and future generations will benefit from my ability to sustain and conserve natural resources.*

*I will support research efforts directed toward more defficient production of a wholesome food supply.*

*I believe it is my responsibility to produce a safe and wholesome product.*

*I believe it is the purpose of food animals to serve mankind and it is the responsibility of all human beings to care for animals in their care.*

Similarly, the National Pork Producers Council has adopted the following Pork Producers Creed:

*I believe in the kind and humane treatment of farm animals and that the most efficient production practices are those that are designed to provide comfort.*

*I believe my livestock operation will be more efficient and profitable if managed in a manner consistent with good husbandry practices as known and recommended by the animal husbandry community.*

*I believe in an open door policy to visitors to my farm, to all those who are sincerely interested in production methods and the welfare of animals, so long as they do not endanger the health and welfare of my animals and do not interrupt my production routine or impair the production process.*

*I believe in and will support research efforts designed to measure stress of farm animals and directed toward more efficient production of food and enhancement of the welfare of animals and man.*

*I believe it is the animal's purpose to serve man; it is man's responsibility to care for the animals in his charge. I will vigorously oppose any legislation or regulatory activity that states or implies interference with that responsibility.*

## Experimentation on Animals

Another issue of concern for animal welfare activists is the use of animals for research. For many years animals have been used to test products and procedures for humans. For example, before any drug is tested on humans, it must be first tested on animals. Medical procedures such as operations and

**FIGURE 18-9**
Some people object to using animals for experimentation.

vaccinations are tested on animals. Most of the medical advances made by humans during the past century have come about through the use of animals to test treatments and medication. Research scientists argue that the use of experimentation on animals is well justified by these advances. They cite the examples of diseases, such as polio, that have been almost eradicated through the use of research using animals in experiments.

Most people realize that research is carried out using mice, rabbits, and guinea pigs. However, during recent years much controversy has come about over the use of cats, dogs, and primates. Some people feel that it is worse to use animals that are "of a higher order." Those who oppose the use of animals in research contend that the knowledge gained through the research cannot justify the suffering the animals must undergo. They advocate the use of computer simulations to test treatments. Researchers point out the computer simulations are helpful, but nothing can ever replace real animals.

Research scientists have a much more difficult time justifying the use of animals to test such products as cosmetics. Activists insist that the suffering of animals should not be allowed merely to produce new products that are used only to make people appear more attractive.

# Animal Rights Activists

Another group concerned with the treatment of animals are animal rights activists. They contend that animals have as much right to enjoy life and to be free as do humans. They believe that killing an animal is just as wrong as killing a human. They oppose the raising of animals and the use of animal products. They also strongly oppose hunting and feel that animals should be free to live in the wild.

Opponents to animal rights activists contend that animals are not the equal of humans and that they exist for use by humans. They argue that if animals were turned loose, both animals and humans would suffer. Animals that have been bred to be cared for by humans could not care for themselves. If animals were turned loose without care they would lead short, miserable lives. Domesticated animals live the best lives under the care of humans.

Opponents also point out that the loss of animal products would be a burden to humans. Nutrients obtained from the eating of meat would be difficult to replace from other sources. The absence of drugs and other pharmaceuticals made from animal

**FIGURE 18-10**
Animals that have been bred to be cared for by humans could not live on their own. *Courtesy of James McNitt.*

by-products would create problems for those people whose lives depend on these products.

Animals left to live on their own would have a difficult time reproducing and surviving. When left entirely on their own in a closed ecological system, animal populations might be able to achieve a balance. However, human populations are large enough to prevent this from happening. A good example is that of deer populations that are allowed to grow entirely on their own. Unless the population is kept under control by measures such as hunting, there are too many of the animals. When this happens, the herds of animals starve or die of disease. A properly controlled population is healthier and the animal population can be sustained.

FIGURE 18-11
Hunting helps prevent wildlife from becoming overpopulated. *Courtesy of Jeff Jackson.*

## Summary

In the future, agriculture will have to contend with an increasing number of people who advocate the ethical treatment of animals. This means that agriculturists will have to make sure that their management practices are proper and that they treat the animals well. In addition, they must make sure that the public is educated about how agricultural animals are treated. As with most issues, communication between the opposing sides is vital. This will help ensure that the ethically correct solution is found.

# 18 CHAPTER REVIEW

## Student Learning Activities

1. Some people feel that rodeos are cruel to animals. Research the topic and draw your own conclusion. List your reasons for making your conclusion. An alternative is to make a conclusion as to whether or not you feel that testing cosmetics on animals is justified. In class, debate others who hold the opposite opinion.

2. Make a list of all the ways animal producers protect their animals. Also make a list of management procedures that might appear to be cruel but are in the best interests of the animals. Be sure to include the reasons why they are beneficial.

3. Visit the local humane society and talk with the employees. With their help, devise a plan for your class to help stop the mistreatment of animals.

## True/False

1. ___ Today in the United States, only a small percentage of the people really understand the raising of farm animals.

2. ___ Most of the animals raised are produced on corporate-owned farms.

3. ___ Pigs are unsociable animals and like to be separated from other pigs.

4. ___ Most animals do better when raised with other animals of their own kind.

5. ___ Animals that are kept in confinement can receive better care than animals that roam free.

6. ___ Animals under the care and protection of humans have no need for horns.

7. ___ Castration keeps animals clean.

8. ___ Hot-iron branding is used less now than in the past.

9. ___ Animal rights activists believe that killing an animal is just as wrong as killing a human being.

10. ___ The National Pork Producers Council has adopted a Pork Producers' Creed.

# Multiple Choice

1. Animal welfare activists disagree with producers over what constitutes proper
   a. neutering.
   b. treatment.
   c. confinement.

2. Producers are in the business of raising animals for
   a. pets.
   b. profits.
   c. insect control.

3. Animals that are placed in crates a week before they give birth are
   a. sows.
   b. cows.
   c. horses.

4. A process performed to help keep animals clean is
   a. neutering.
   b. debeaking.
   c. docking.

5. An issue of great concern for animal welfare activists is the use of animals for
   a. food.
   b. pets.
   c. research.

6. Before any drug is tested on humans, it must first be tested on
   a. insects.
   b. animals.
   c. bacteria.

7. Animals that live the best lives under the care of humans are called
   a. confined.
   b. domesticated.
   c. wild.

8. An example of a disease that has almost been eradicated through the use of research is
   a. polio.
   b. cancer.
   c. diabetes.

9. Organisms that cause stress and discomfort to animals and may even shorten their lives are called
   a. herbivores.
   b. parasites.
   c. sows.

10. A procedure that makes animals more docile and less aggressive is
    a. neutering.
    b. dehorning.
    c. confining.

# Discussion

1. What are two ways of thinking concerning the well-being of animals?

2. Why is it important that animals have pleasant surroundings?

3. Why is stress so detrimental to animals?

4. How do animals benefit from confinement?

5. What management practices do animal welfare activists disagree with?

6. How have scientists justified the use of animals for research?

7. Why do scientists have a more difficult time justifying the use of animals to test cosmetic products?

8. What are the beliefs of animal rights activists?

9. Explain how deer are a good example of the human impact on controlling animal populations.

10. What is the National Cattlemen's Association statement of principles based on?

# Biotechnology: The Future of Agriculture

KEY TERMS

biotechnology
artificial insemination
embryo transfer
clone
genetic engineering

## Student Objectives

*When you have finished studying this chapter, you should be able to:*

- Relate why agriculture will always exist.

- Cite examples of how biotechnology has been used in the past.

- Explain the current uses of genetic engineering.

- Discuss some future uses of genetic engineering.

- Explain the concerns over genetic engineering.

- Discuss careers in agricultural science.

Over the years many industries and occupations have come and gone. As jobs outlive their usefulness, the industry dies out. One industry that can never die out is agriculture. People must have food, shelter, and clothing. All of these needs, especially food, depend on agriculture.

FIGURE 19-1
Since ancient times many changes have occurred in the way humans grow food. *Courtesy of James Strawser, The University of Georgia.*

Even though agriculture will always exist, the way things are done in agriculture will change. Just look at all of the changes that have occurred since humans first began to till the earth using sharpened sticks. Although changes took place over thousands of years, most have occurred during the last century. Futurists tell us that the changes you will see during your lifetime will be greater than any that have occurred so far. That is hard to imagine!

# Biotechnology

The most profound changes in the future will occur in **biotechnology** This term refers to the application of biological and engineering techniques to living organisms. This concept is not new. In fact, biotechnology has been in use for thousands of years. Perhaps the earliest biotechnology was used to make cheese and wine. Grape juice and milk spoiled very rapidly. Through the use of certain types of bacteria, the milk could be fermented to form cheese and the juice could be fermented to form wine. The use of these bacterial agents transformed highly spoilable foods into foods that could

FIGURE 19-2
Long ago grape juice would spoil unless it was fermented into wine. *Courtesy of James Strawser, The University of Georgia.*

be kept for a long period of time. In the same manner, food such as fresh cabbage could be fermented to make sauerkraut that would last longer.

Ancient biotechnology was not limited to food preservation. Legend has it that hundreds of years ago, Arabs first used **artificial insemination.** These people were noted for the beautiful horses they grew. According to the legend, Arabs from one tribe would sneak into the camp of enemy tribes and collect semen from the prized stallions of the enemy chieftains. They would then use the semen to impregnate their mares. This artificial mating produced fine horses.

## Reproductive Technology

Reproductive technology has progressed dramatically since the time when the Arabs first began artificial insemination. The use of both artificial insemination and **embryo transfer** are now commonplace technologies. Both dams and sires are chosen through the use of computers. The ability to freeze and store semen and embryos has tremendously affected the production of superior animals.

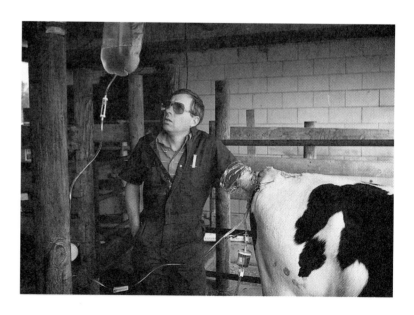

**FIGURE 19-3**
The use of artificial insemination and embryo transfer have greatly advanced the animal industry.
*Courtesy of James Strawser, The University of Georgia.*

Scientists can now even **clone** animals. Cloning is the ability to produce an organism that is genetically identical to another. Plant scientists have been able to clone plants for sometime using a technique called tissue culture. Animals have been more difficult to clone. However, several companies now offer frozen embryos for sale that have been cloned from superior parents.

Another reproductive technique that is just around the corner is the development of all-male or all-female embryos. This technology would be valuable to producers who only produce one or the other sex. For example, if a producer is in the business of hatching pullets to be grown into laying hens, only the females are valuable. Now, about half of the eggs that hatch are male. If the process could be developed to produce only eggs that would hatch female chicks, the efficiency of the hatchery could be doubled. Another example would be the production of heifers for dairies. Bull calves are not nearly as valuable as female calves that will one day produce milk. The technology already exists to produce embryos that are only male or only female. The challenge is to make the process economically feasible.

## Genetic Engineering

The newest form of biotechnology is **genetic engineering.** Genetic engineering is the altering of the genetic makeup of organisms. Genes are the substances that determine the characteristics of an individual organism. One of the first uses of this new technology was the manufacture of human insulin. This substance is produced by the human body and helps to regulate the metabolism of carbohydrates. If the body does not manufacture enough insulin, a condition known as diabetes results. Previously, insulin to treat diabetes was harvested from the

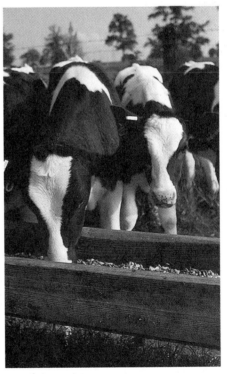

**FIGURE 19-4**
Dairy producers place a much greater value on female calves than they do on male calves. *Courtesy of James Strawser, The University of Georgia.*

pancreases of slaughtered cattle. Since only a tiny amount could be obtained from an animal, the process was expensive. Also, cattle insulin was not exactly the same as human insulin and some people were allergic to it.

This problem was solved through the use of genetic engineering. Scientists were able to find the

## *Career Development Events*

**FFA**

The FFA Agriscience student recognition program not only rewards the students, but it is designed to educate parents, school personnel, and the public about the career opportunities available in agriscience. The program recognizes students who are pursuing challenging school-based courses focused on the scientific principles and new technologies of agriculture. The winners of this award receive financial assistance to attend college. This helps to produce a reliable supply of agriscience graduates to meet society's future needs.

To be eligible for participation in the agriscience recognition program, the student must be an FFA member who is a high school junior or senior or a first-year student in college. The student must be taking academically demanding courses including classes in agriculture and be planning a career in agriculture

that requires at least a bachelor's degree. The student must submit an application and list of agriscience activities and completed projects. These activities will be judged by awarding points for skill, scientific thought, thoroughness, clarity, and creative ability. The student's academic achievement and school and community activities are also judged.

The student's agriscience project must follow the scientific research process. Each phase of the project must be identified, explained, and recorded. Tables, lists, or graphs are a good way to illustrate observations made during the experiment. The agriscience project should fall into a plant science, animal science, natural resources, earth science, energy production and management, food science, or engineering science categories. Applications should include letters of recommendation from an employer, instructor, counselor, or other school official plus pictures showing the student participating in the project.

gene part that regulates the production of insulin. They then spliced this segment into the gene of the *E. coli* bacteria. The bacteria that produce insulin multiply and pass the capability along to the next generation. Since the genetically altered bacteria are weaker than the ordinary bacteria, they have to be grown under carefully controlled conditions.

The Agriscience Student Recognition Program must be submitted through the agriculture instructor. The state winner will be presented a scholarship for use at the college of choice and is eligible to compete for regional and national awards. Regional and national winners can win scholarship money that totals over $8,000.

The Agriscience Award recognizes students who have excelled in the science of agriculture. *Courtesy of Blane Marable, Morgan Co. High School, Madison, GA.*

When the proper number of bacteria are reproduced, they are taken apart to retrieve the insulin produced by the bacteria. The insulin is purified, and the remains of the bacteria are destroyed. The genetically altered bacteria produce a ready, relatively inexpensive supply of insulin.

Bacteria have become the manufacturing centers for many substances that have made the lives of humans better and more productive. Vaccines that are used to impart immunity into animals and humans are produced in much the same way as insulin. Hormones that control growth and other bodily functions are also produced using genetic engineering.

Through genetic engineering, scientists have developed plants that are resistant to herbicides. Herbicides are chemicals that are used to kill weeds or unwanted plants. When the weeds are growing among a crop, a herbicide must be used that will kill the weeds and not harm the crop. Genetic engineering can develop crop plants that are tolerant of herbicides. This is a tricky process because the genetically manipulated crop cannot be guaranteed to retain all of the characteristics of the nonmanipulated crop. For example, grain that is grown for food still has to produce well and also must contain the proper amounts of nutrients.

Scientists have developed the ability to determine exactly where the part of the gene is located that controls certain characteristics. They can do this for some characteristics on some organisms. Just think of the possibilities when they are able to map out and identify exactly where every characteristic is located for every organism. For example, bacteria live in a ruminant's stomach and allow the animal to digest large amounts of fiber. One day scientists may be able to genetically alter the bacteria to make them able to digest wood fiber. This would allow cattle and other animals to be fed on woody fiber that is now going to waste. Perhaps monogastric (simple-stomached animals) such as pigs could be developed that could digest roughages like ruminants. This would mean a much cheaper way to produce pork.

Another possibility may be to develop plants like corn that produce their own nitrogen. Corn requires a lot of nitrogen in order to produce large yields. This fertilizer is expensive to supply. Some plants, called legumes, produce nitrogen through the use of bacteria that live in nodules on their roots. If scientists could genetically alter the corn plant (along with other grasses) to allow them to host the nitrogen-producing bacteria, grain could be produced much more efficiently.

Crops may also be altered to produce more oil or oil of a better type. Remember from Chapter 10 that some crops are produced for the oil from their seed. Scientists think that in the future these as well as other crops may be altered to produce a much higher percentage of oil. This may go a long way in helping to solve the problems caused by running out of petroleum.

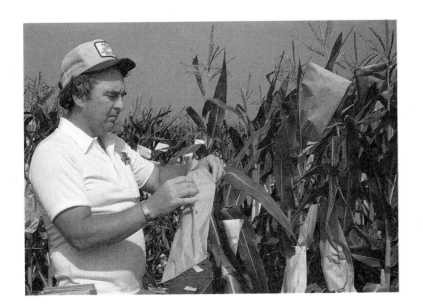

**FIGURE 19-5**
If scientists could alter corn to make it a legume, much less would be spent on fertilizer. *Courtesy of Dekalb Genetics Corporation.*

Genetic engineering has the potential to be revolutionary in the production of food and fiber. Scientists estimate that by the year 2000 there will be over 6 billion people in the world. All of these people will have to be fed and clothed. Through the use of genetically altered plants and animals, the efficiency of producing food can be increased and areas that are currently of little agricultural value can be made productive.

Even though there is much hope and optimism over genetic engineering, there is also widespread concern. Some people are worried about the introduction of genetically altered organisms into the environment. They cite examples of plants and animals that were introduced with good intentions and today have escaped and caused problems. They feel that by tampering with nature we may unleash organisms that will be hard to control. Many science fiction movies depict scary creatures that have been genetically altered. People note that the most commonly used organism in gene alteration is the *E. coli* bacteria that normally live in the human intestinal tract. Since the bacteria multiply rapidly and the human body is a host for these bacteria, many people feel that we do not know enough about genetic alteration to allow these bacteria into the environment.

The answer to this dilemma is, at least in part, the regulation of genetic engineering. Scientists conduct research on genetic engineering in closely controlled laboratory situations. The buildings and equipment are designed to prevent the escape of newly engineered microorganisms into the environment. Research may take many years before the determination is made that the organisms will be tested outside the laboratory.

The United States Department of Agriculture (USDA) has developed strict guidelines for testing genetically altered organisms that can come in con-

**FIGURE 19-6**

Facilities used in genetic engineering are designed to prevent the escape of organisms. *Courtesy of USDA ARS.*

tact with the environment. In addition, the USDA has developed a set of guidelines that all USDA research agencies will use prior to field testing. A heightened public awareness helps to ensure that those doing research remain diligent and do not become careless. It also helps everyone remain conscious of the fact that problems could result from carelessly conducted genetic research.

## Careers in Agriscience

You may not have thought about what sort of work you want to do, but sooner or later you must decide. Consider a career in the nation's largest industry—agriculture. From stories in the media, you may think that agriculture is dying out. Nothing could be further from the truth. Agriculture is not dying—it is changing. The changes are exciting and challenging. Remember that all the people in the world will have to be fed and clothed. This will be done through agricultural science.

Consider all the things you like to do. A substantial part of your life will be spent working, and if you enjoy your work, your life will be more

**FIGURE 19-7**
Perhaps you should consider a career in the nation's largest industry. *Courtesy of Patrick Smith, The University of Georgia.*

**FIGURE 19-8**
Agriculturists work in a variety of settings. *Courtesy of USDA ARS.*

pleasant. If you like to work with plants or animals and if you like science, a career in agriculture might be for you. Some people have the mistaken idea that a career in agriculture is difficult, strenuous, dirty work that has no challenge. Today's agriculture can offer exciting challenges. You may work in a variety of places ranging from a food processing plant to a laboratory.

## Summary

Like any other career, you will need the proper preparation for a career in agriculture. The best place to begin is with the high school agricultural education program. In this program you will decide which areas you wish to specialize in. Throughout this book are examples of activities you can get involved in through the FFA. The opportunities are limitless. You can even earn scholarships to help pay for a college education. Agriculture is taught as a two-year program in community and technical colleges, or as part of a four-year university program. These institutions will be glad to send information on opportunities in all their degrees and programs.

# 19 CHAPTER REVIEW

## Student Learning Activities

1. Think of a problem facing agriculture today or in the future. Write a paper giving your ideas as to how the problem could be solved using biotechnology. Be sure to include possible hazards.

2. Interview five people and get their perceptions of genetic engineering. Do you feel their perceptions are correct? Report to the class on your conclusions. Be sure to keep the identity of the people you interviewed confidential.

3. Research the career opportunities in biotechnology. Determine the job responsibilities of a particular career and the qualifications needed. Report to the class on your findings.

## True/False

1. ___ Most changes in agriculture occurred during the last century.

2. ___ Dams and sires are chosen through the use of computers.

3. ___ Scientists have not yet discovered how to clone animals.

4. ___ The majority of insulin is harvested from the pancreases of cattle.

5. ___ Corn requires a lot of nitrogen in order to produce large yields.

6. ___ Scientists estimate that by the year 2000 there will be over 6 billion people in the world.

7. ___ Legumes produce nitrogen through the use of bacteria that live in nodules on leaves.

8. ___ The nation's largest industry is agriculture.

9. ___ Agriculture is not dying; it is changing.

10. ___ A career in agriculture may involve working in a variety of places like a food processing plant or a laboratory.

# Multiple Choice

1. One industry that can never die out is
    a. agriculture.
    b. computer production.
    c. engineering.

2. Perhaps the earliest biotechnology was used to make wine and
    a. bread.
    b. apples.
    c. cheese.

3. One of the newest forms of biotechnology is
    a. insecticide production.
    b. manufacturing herbicides.
    c. genetic engineering.

4. One of the first uses of genetic engineering was in manufacturing
    a. cotton.
    b. human insulin.
    c. pesticides.

5. Genetic scientists have been able to develop plants that are resistant to
    a. herbicides.
    b. biotechnology.
    c. predators.

6. Bacteria live in a ruminant's stomach and allow the animal to digest large amounts of
    a. fiber.
    b. biotechnology.
    c. grain.

7. The best place to begin preparing for an agricultural career is with an agricultural education program beginning in
   a. preschool.
   b. elementary school.
   c. high school.

8. The most profound changes will occur in
   a. building construction.
   b. genetic engineering.
   c. animal care.

9. Food, shelter, and clothing are all directly dependent on
   a. agriculture.
   b. crops.
   c. water.

10. Substances that control growth and other bodily functions are
    a. vaccines.
    b. hormones.
    c. herbicides.

# Discussion

1. Why is cattle insulin no longer used to help diabetics?

2. Why do scientists want to genetically alter the bacteria found in a ruminant's stomach?

3. What impact does genetic engineering have on plants?

4. Why is there concern about genetic engineering?

5. How is genetic engineering being regulated?

6. How can you prepare for a career in agriculture?

7. Why will agriculture always exist?

8. What are some current uses of genetic engineering?

9. What are some future uses of genetic engineering?

10. How was biotechnology used in the past?

# Careers in Agricultural Science

## Student Objectives

**When you have finished studying this chapter, you should be able to:**

- Identify the main job classifications in agriculture.

- Analyze the career opportunities in agriculture.

- Identify some of the characteristics associated with various jobs in agriculture.

D uring the first half of the 20th century agriculture was mostly limited to farming, and preparing for a career in agriculture was fairly easy. People did not think that a formal education was necessary for a farmer. Today, successful people who work in agriculture need a good education in business and in science. Over the years, the number of farmers in the United States has declined dramatically, yet agriculture remains a major source of

*This chapter was contributed by Jean Kilnoski.

jobs for Americans. Agriculture professionals are not just farmers and ranchers. Opportunities in this field include areas as diverse as science, engineering, finance, marketing, and exporting. Careers in these areas have expanded as the production and processing of food and fiber have become more specialized and technological. Successful producers rely heavily on the support of these workers. Individuals working in agriculture today are as likely to work in the city as on the farm.

The business of agriculture, or **agribusiness,** is an important part of the U.S. economy. More than 20 million workers are employed in processing, storing, and distributing the food and fiber produced by farmers. Agribusiness also involves manufacturing and selling farm supplies and equipment. The continued success of agriculture depends on the support of this entire network of people who enable the producer to function more efficiently. It is their initiative and hard work that make American agriculture competitive in the world market.

**FIGURE 20-1**
More than 20 million workers are employed in processing, storing, and distributing food and fiber. *Courtesy of Henry Waeti.*

## Plant Science

Plant science deals with the growing of fruits, vegetables, and ornamental plants. Career opportunities in plant science vary widely, but all of them are involved in producing plants. The main categories in plant science are horticulture, agronomy, landscaping, and turf management.

The production of field crops requires the expert knowledge of many scientists. **Agronomists** provide producers with information about the combination of soils, plants, and climate that will promote the best possible yield. **Entomologists** provide information about pest control. Today insect populations are not always controlled with pesticides. Parasitic insects can be released in a field to control a pest population. Crop scientists work to increase

**FIGURE 20-2**
Crop scientists work to increase the yield of crops by improving new plant strains. *Courtesy of USDA.*

the yield of crops by improving farming methods and developing new plant strains. Some crop scientists study the breeding of crops and use genetic engineering to develop crops resistant to pests and drought. All of these agricultural scientists must have advanced college study in their chosen fields.

Horticulture deals with the cultivation of fruits, vegetables, flowers, trees and shrubs. Some **horticulturists** find new ways to grow plants or breed better varieties. They experiment with different varieties of fruits, vegetables, and nuts to make them more nutritious. Others raise and tend plants used for ornamental purposes, such as flowers, shrubs, trees, and grasses. An increase in the interest for indoor and outdoor plantings has raised the demand for these services. People want attractive grounds around their homes and offices, as well as amusement parks, gas stations, fast-food stores, and shopping malls. Skilled horticulturists are needed to grow and tend plants in all of these areas.

Many jobs in plant science require two years of study or less. Nursery operators raise and sell flowers, shrubs, hedges, and trees. They are involved in tasks including plant propagation through seeds and cuttings, preparation of soil in outdoor growing areas, greenhouse management, and the storage and packing of plants.

Landscape architects design outdoor areas for people to use, such as parks or gardens, housing projects, or the grounds surrounding hotels. They plan their arrangements to make the best use of the land while protecting the natural environment. They prepare detailed maps and plans showing the placement of trees, shrubs, and walkways. The landscape architect also supervises any grading or construction that is necessary to complete the work.

Turf grass technicians may work in commercial lawn maintenance, private lawn care, or the maintenance of parks, highways, and playing fields. They may be employed as greens keepers for golf courses, work for the government maintaining public lands and parks, or may work in their own businesses. They provide numerous services that include mowing and fertilizing, as well as insect, weed, and disease control.

**FIGURE 20-3**
Many careers in turf and landscaping require less than a four-year college degree. *Courtesy of* Progress Farmer Magazine.

## Animal Science

Careers in animal science are involved with many phases of animal production. These professions include producers, feeders, marketers, manufacturers, distributors, and retailers. They may also include producing pharmaceuticals or equipment that aids in the management of animals. Many of these careers are highly science oriented. The intensive health practices that are necessary to maintain a

## Career Development Events

**FFA**

The National FFA Convention was founded in 1928, and every year since that time FFA members from all over the country have met in Kansas City. The convention has grown to become the world's largest student educational convention. Students from every state in the country, plus students from several foreign countries, attend. All types of career development events are conducted during the convention ranging from public speaking to agriscience. Tours to educational sites, such as the Truman Library, are conducted almost every day. Students have a wide array of leadership development workshops to choose from where they can interact with students from all over the country.

At the convention, students receive awards for many activities that began on the local level and progressed through to the national level. It is quite a good feeling to know that they have earned a national award through hard work and dedication. The FFA awards over a million dollars in scholarships each year.

One of the many highlights of the convention is the gigantic Careers Show. Companies from all over the world set up booths to tell about their company. Students can learn about thousands of job opportunities involved with agriculture. Many colleges from all across the United States operate booths where students can get information about various programs and scholarships. They can go home with stacks of printed materials about job opportunities.

Most of the students attending the

successful operation in cattle ranching or dairy farming provide many job opportunities in animal welfare. Scientists study the nutrition, growth, and development of domestic animals. **Geneticists** look for ways to develop high growth-rate animals that result in a quality product that meets consumer standards. These positions require a four-year program of college study that emphasizes the sciences.

Cattle reproduction has opened the door to another career in the animal sciences. Artificial in-

convention have earned their way. As you begin to study agriculture in high school, set your goal to attend the National FFA convention. Your teacher can help you design a plan to achieve your goal. Remember, the opportunities in FFA are limited only by your ambition and willingness to set goals and work toward them.

The National Career Show at the National FFA Convention provides information for students on hundreds of different careers in agriculture. *Courtesy of Bill Stagg, National FFA Organization.*

**FIGURE 20-4**
Animal scientists study the nutrition, growth, and development of animals. *Courtesy of The University of Minnesota Agricultural Experiment Station.*

semination is the most common cattle breeding practice today. Artificial insemination technicians typically have two years of college education. Feedlots also present career opportunities in animal science. At the feedlot the cattle are fattened up so that their market value improves. The manager of a feedlot is responsible for minimizing the costs of feeding the cattle in order to obtain the highest profit. Feedlot operations use computers to regulate feed supplements and medical products. Many feedlot managers are producers who grow their own hay and grain.

Cooperative extension agents provide services to both producers and apartment dwellers. Agents are able to advise people on topics as diverse as crop and livestock production, conservation, home economics, child care, farm management, and nutrition. Cooperative extension agents can answer questions about when to transplant seedling tomatoes or how to remove stains from clothing. Agents often specialize in one of these areas. Extension agents are employed jointly by county and state governments. They must have a four-year degree in one of the fields that relates to extension work.

If you enjoy being around animals you might want to consider pursuing a career in one of the many areas dealing with companion animals. The most important attributes you need for a career in companion animal care are respect and love for animals.

Pet care workers provide a wide variety of services to the owners of small animals. They work in animal hospitals, boarding kennels, and animal shelters. Responsibilities will vary, but may include feeding animals and cleaning and disinfecting cages and equipment. Pet care workers can also work in pet grooming parlors, pet training schools, and pet shops. Many of these positions are entry level and require no special education. Employers often hire high school students and graduates to fill these positions. Some experience with animals is helpful.

Animal trainers teach animals to obey commands, compete in shows or races, or perform tricks. Many companion animals can be trained, including dogs, horses, parakeets, and cockatoos. Animal trainers can teach dogs to perform a variety of tasks for humans, including the protection of property. Employees in this field may also be involved in training assistance dogs for the blind, the deaf, or the physically challenged. Most animal trainers begin their careers as pet care workers where they gain experience caring for animals.

**FIGURE 20-5**
People who train animals are also in an agriculturally related field. *Courtesy of Kentucky Horse Park.*

Dog groomers wash, comb, cut, and trim the fur of dogs and cats. To become a dog groomer it is best to take a course from one of the 50 schools accredited by the National Dog Groomers Association. To enroll in the school, you must be at least 17 years old. Some animals may be nervous, fidgety or aggressive, so patience and persistence are necessary characteristics of a potential dog groomer.

Horse industry workers breed, train, and care for horses. People who work with racehorses usually begin their careers as grooms. Grooms are responsible for brushing and combing the horses. They are often in charge of feeding the animals and cleaning the stables where they are kept. Grooms may also be involved in training young horses.

Horse trainers are responsible for determining the best diet and exercise program for the horses. The trainers often supervise grooms and other employees who exercise and care for the horses. Nearly all horse industry workers learn at least some of their skills on the job.

Animal health technologists work side-by-side with veterinarians. They assist veterinarians with laboratory work, radiology, anesthesiology, and surgery. They may also be responsible for animal nursing, sample collection, and general office management. An animal health technologist is not limited to work in a veterinary clinic or laboratory; he or she may also work in a zoo or a circus. These technicians are graduates of a two-year program from a community or technical college.

Veterinarians are animal doctors. They perform a variety of duties in a number of different environments. Veterinarians diagnose animal illnesses, treat diseased and injured animals, perform surgery, and prescribe and administer drugs and vaccines. They also give animal owners advice on the care and breeding of animals. They can work in a clinic, conduct research for the government, or work for a

**FIGURE 20-6**
Animal health technologists may assist with laboratory work. *Courtesy of USDA.*

large organization. They can treat both large and small animals, or they can specialize. Almost one-third of all veterinarians specialize in the care of small animals.

Veterinarians must complete four years of college followed by an additional four years of study at a veterinary school. To become licensed to practice, veterinarians must also pass their state's oral and written licensing examination.

# Food Science

The food industry is one of the largest industries in the nation. It offers a wide range of opportunities because of the variety of operations and services it performs. Only a small portion of the foods we consume are fresh. Most have been processed by the time they reach the local supermarket. Before the food product can be sold to the consumer it must be processed, inspected, packaged, labeled, stored, and transported. All of these jobs are part of the extensive processing of food products, from the time the crop is harvested until it reaches the consumer.

One of the main purposes of processing foods is to keep them from spoiling. Processed foods are canned, waxed, dried, frozen, bottled, and pickled to preserve them. Another purpose for processing is to enhance the nutritional value of the food. Highly technical methods of achieving these goals have been developed that require the skills and knowledge of **food scientists, microbiologists,** and food chemists. All of these positions require advanced college study.

Food scientists improve methods of freezing, canning, storing, packaging, and distributing food. They study the chemical changes that take place in food when it is stored or processed. Food scientists try to find ways to process food so that fewer

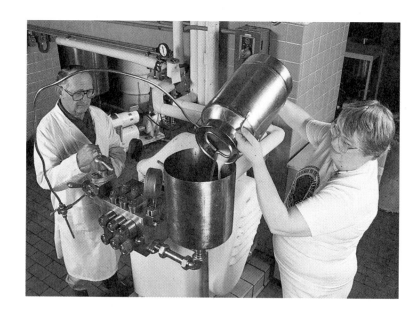

**FIGURE 20-7**
Food scientists improve methods of storing, packaging and distributing food. *Courtesy of USDA.*

nutrients are lost. They also study the effect of food additives. Many food scientists are attracted to research and development. The most thrilling task of the food scientist is the development of new products. This is a highly competitive area. Developing a product usually takes more than a year and a substantial financial investment by the processing company. Only one product out of ten will survive the first year in the supermarket. Ready-to-cook convenience foods continue to account for a major part of the new product development efforts. Thanks to convenience foods, people are able to spend significantly less time in the kitchen preparing meals than they did just a decade ago. Frozen, precooked meals that come in microwaveable packages are a good example of the innovations that have occurred in the food industry in recent years.

Careers in meat processing often deal with the protection of processed meats. Microbiologists lead the way in food safety research. They help the industry find ways to prevent bacterial growth. Food chemists work in the meat industry to ensure a healthful, tasty product. They may study the chemistry of fats in chicken skins to help reduce the

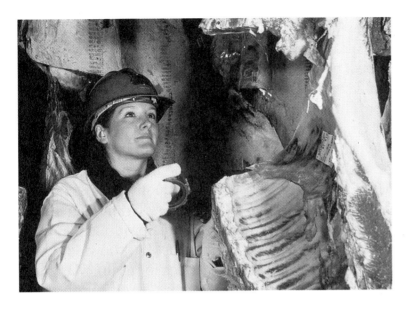

**FIGURE 20-8**
Meat inspection is a career that focuses on a safe product for consumers. *Courtesy of USDA.*

cholesterol in human diets or the muscle chemistry of meats that causes some cuts to be tender and others to be tough. Meat inspection is another career that focuses on a safe product for the consumer, whether it is poultry, red meat, or fish. Inspectors who work for the Department of Agriculture inspect and grade almost all commercially marketed meats.

Food technicians graduate from a two-year program of study. They work in food processing plants and in laboratories. They usually help with quality control, testing or research under the supervision of a food scientist. Technicians run tests to check processed foods for bacteria, impurities, or poisons. They use instruments that check the quality of the food, including the taste, smell, and color. Some food technicians work as production supervisors and managers in food processing plants.

A number of careers in the food industry deal with agricultural marketing. Agricultural marketing involves the storing, transporting, or marketing of food products. Supermarkets are dependent on the efficient functioning of the agricultural marketing sector. Transportation specialists use all of the forms of transportation available, trucks, trains, and barges, to get the lowest price possible while maintaining the quality of the product. These specialists must have a good working knowledge of the state and federal laws and regulations that govern the transportation of these products.

Grain merchants are involved in all aspects of buying and selling grain. They act as a connection between the producer and the ultimate consumer. They are concerned with the quality, the shipping, and the processing of the grain. Grain merchants buy most of their grain from area producers at local grain elevators. Merchants must be knowledgeable about the current market prices for grain. Many use computers to get up-to-date prices. Although there are no specific education requirements for grain merchants, some college training is helpful.

The wholesaler is a link in the marketing chain between the producers and the retail store in which the goods will be sold. The wholesaler provides a valuable service by bringing these two segments of the marketplace together. Wholesalers also buy goods in the United States for sale abroad, as well as buying products from foreign markets for sale here.

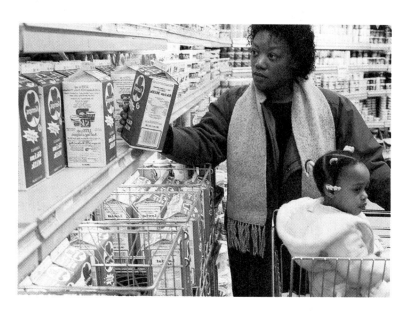

**FIGURE 20-9**
Supermarket managers can be responsible for more than 15,000 food and consumer items. *Courtesy of The University of Minnesota Agricultural Experiment Station.*

The retail store manager who buys the food product from the wholesaler is responsible for every phase of the store's operation. Supermarket managers can be responsible for more than 15,000 food and consumer items. They also supervise other employees, manage the inventory, account for expenses, and maintain a good relationship with the consumer. The retail manager must constantly evaluate how well each product is doing in order to make effective decisions about when to eliminate one product from the shelf to make room for a new one. This type of evaluation is often made possible by electronic devices that read product bar codes. There are no specified educational requirements for wholesalers and retailers, but a college education is highly recommended.

## Natural Resources

In recent years, people in agriculture have been challenged to find ways to grow enough crops to satisfy the needs of consumers without damaging the environment. This problem will become more difficult to solve as the demand for food increases. Many people in agriculture are devoted to the protection and conservation of the environment. Highly educated and trained people will be needed to protect the nation's forests, wilderness areas, and recreations areas. Most of the careers available in this area require a four-year college degree.

Soil scientists study the characteristics of soil. They gather information and give advice on the management and conservation of soil and water as well as erosion. Many soil scientists work for the United States government. These scientists are called soil conservationists.

Range managers oversee the use of rangelands that have been set aside for livestock grazing and

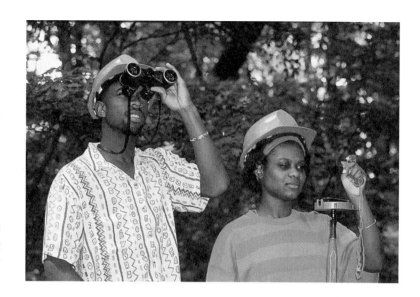

**FIGURE 20-10**
A career with the forestry service may deal with protecting the environment. *Courtesy of USDA.*

**FIGURE 20-11**
Agricultural research will become even more important in the years to come. *Courtesy of USDA.*

other uses. They supervise the use and conservation of rangelands owned by the government. Range managers balance the grazing needs of ranchers and conservation measures needed to protect the environment. The majority of range managers work for the government. Others may work as appraisers for banks or for companies that have large holdings of rangeland.

Another career that deals with protecting the environment is with the U.S. Forest Service. Forest Service researchers conduct studies on wildlife and fish. They study the habitat requirements of individual species to determine how certain activities in the forests and on the rangelands will affect them. The studies allow them to make informed decisions about cutting timber or allowing grazing of the rangelands.

## Summary

The future of careers in agriculture looks bright. If current predictions hold true, workers will need to become even more efficient because the demand for food and other agricultural goods will continue to increase. If the world's growing population is to be fed, world food production must double in the next 40 years. In addition, changing consumer tastes will continue to prompt new product development. Agricultural research will become more important in the years to come as new technologies are needed to meet the needs of the American consumer while maintaining a healthy environment.

# 20 CHAPTER REVIEW

## Student Learning Activities

1. Choose a career that interests you. List as many of the tasks performed in this career as you can. Find out what qualifications are required to be hired in this position.

2. Interview a person who works in the career you have chosen. List all the things he or she enjoys about working in that job.

## True/False

1. ___ Agriculture is the same thing as farming.

2. ___ Horticulturists may raise plants for ornamental purposes.

3. ___ Cooperative extension agents only serve producers.

4. ___ Most of the foods that we consume today are fresh.

5. ___ Careers in meat processing often deal with the protection of processed meats.

6. ___ One of the main purposes of processing foods is to alter the flavor.

7. ___ There will be fewer jobs in agriculture in the future.

8. ___ An individual looking for a job in agriculture today will probably work in the country.

9. ___ More than 20 million people are employed in agribusiness.

10. ___ Almost one-third of veterinarians specialize in the care of small animals.

# Multiple Choice

1. The business of agriculture is called
    a. horticulture.
    b. agronomy.
    c. agribusiness.

2. Entomologists provide producers with information about
    a. pest control.
    b. the best combination of soils, plants, and climate.
    c. breeding ornamental plants.

3. Individuals who deal with the chemistry of fats in chicken skins are
    a. food chemists.
    b. microbiologists.
    c. entomologists.

4. Horticulturists are experts in
    a. designing outdoor areas for people to use.
    b. how to grow, pick, and ship plants.
    c. artificial insemination.

5. Individuals who give advice on the conservation of soil and water are
    a. soil scientists.
    b. agronomists.
    c. Forest Service researchers.

6. One of the purposes of processing food is
    a. to decrease the nutritional value.
    b. to study food additives.
    c. to keep it from spoiling.

7. A major part of new product development efforts are due to
    a. increased job opportunities.
    b. convenience foods.
    c. the study of food additives.

8. Individuals who act as a link in the marketing chain between the producers of goods and the retailer are
    a. food technicians.
    b. retail store managers.
    c. wholesalers.

9. A landscape architect would be involved in
    a. designing the grounds surrounding an apartment complex.
    b. mowing and fertilizing the grounds of a golf course.
    c. the cultivation of plants.

10. Job opportunities in agribusiness have increased because of
    a. the increase in the number of farmers in agriculture.
    b. an increase in the amount of land that is planted.
    c. an increase in the specialization of the processing of food and fiber.

# Discussion

1. How has the role of education changed for farmers in the last 50 years?

2. How does agriculture continue to be an important part of the economy if the number of farmers has declined?

3. List three jobs in the area of plant science.

4. Why do producers today depend on a network of people?

5. Name three topics about which a cooperative extension agent would be able to answer questions.

6. What are some of the tasks of a food scientist?

7. What is agricultural marketing?

8. What challenge does the environment present to producers?

9. Why does it look like careers in the field of agriculture will continue to be important in the future?

10. Why is food processed?

# GLOSSARY

## A

**A horizon**   The leached upper portion of a soil profile; the eluvial layer.

**Adventitious root**   A root growing from a stem or branch and not from root tissue.

**Aerator**   A mechanical device used to put oxygen into a pond or lake.

**Aerial root**   A root growing from the stem which is exposed to the air.

**Agriculture**   The industry engaged in the production of plants and animals for food and fiber, the provision of agricultural supplies and services, and the processing, marketing, and distribution of agricultural products.

**Alluvial soil**   Soil that is transported and deposited by running water.

**Amino acids**   The building blocks of protein.

**Anaerobic bacteria**   Bacteria that thrive in the absence of oxygen.

**Antibody**   A substance manufactured by an animal's body to combat a disease.

**Animal rights**   The philosophy that supports the idea that animals have the same rights as humans.

**Animal welfare**   The philosophy that supports the idea that animals should be well cared for and well treated.

**Anther**   The part of the stamen on seed-producing plants which develops and contains the pollen.

**Aquatic animal**   An animal that lives in water.

**Aquaculture**   The growing of aquatic animals for human use.

**Aquifer**   A geologic formation in rock or sand that contains water.

**Artificial insemination (AI)**   The placing of sperm in the reproductive tract of the female by means other than the natural breeding process.

**Asexual reproduction**   The propagating of plants from a single plant.

## B

**B horizon**   The mineral horizon below an A or E horizon.

**Bedding plant**   Flower and vegetable plants that are planted in beds.

**Beef**   The meat obtained from cattle. Usually refers to meat from an animal over the age of one year.

**Biological control**   The control of pests using natural means such as insect predators or weed eating insects.

**Biologist**   A person who studies living organisms as a career.

**Biology**   The field of study dealing with living organisms.

**Biotechnology**   The application of biological and engineering techniques to living organisms.

**Bovine somatotropin (BST)**   A naturally occurring hormone that aids in stimulating milk production in cows.

**Brackish**   Water that is neither salt nor fresh but contains a low level of salt.

**Breed**   Animals having a common origin and distinguishing characteristics.

**Broiler**   A chicken raised for market that is 2.5 pounds or more. They are usually 6 to 8 weeks old.

**Brooder**   A device used to keep chicks warm.

**Brood pond**   A pond used to keep fish that breed and lay eggs.

**Bulb**   The below-ground bud of a plant.

**Bull**   A male bovine that has not been castrated.

**Butter**   A product made from milk fat.

# C

**C horizon**   The mineral horizon or layer, excluding indurated bedrock, that is little affected by soil-forming processes and does not have the properties typical of the A or B horizons.

**Caloric content**   The amount of energy a food contains.

**Cannibalism**   The eating or killing of animals by their own kind.

**Canopy**   The uppermost growth of crowns of trees.

**Carcass**   The part of a meat animal remaining after slaughter and the hide, head, feet, and internal organs have been removed.

**Carcinogen**   Any cancer-causing agent.

**Cereal**   Pertaining to grains or products made from grains.

**Cheese**   Food made from the consolidated curds that have been separated from milk.

**Cholesterol**   A fat-soluble substance found in the fat, liver, nervous system, and other areas of an animal's body. It plays an important role in the synthesis of bile, sex hormones, and vitamin D.

**Clay**   A sediment of soft, plastic consistency composed primarily of fine-grained minerals.

**Climax vegetation**   The plants that are the last to appear as a result of natural succession.

**Closed-loop system**   A system of tanks where fish are raised. Oxygen is provided and the water is filtered.

**Cocoon**   The covering that surrounds an insect pupa.

**Cold frame**   An enclosed unheated frame used for growing plants.

**Combine**   A self-propelled or tractor-drawn machine that cuts, threshes, and cleans the standing crop while moving across the field.

**Commodity**   A transportable resource product with commercial value; all resource products that are articles of commerce.

**Concentrate**   A feed high in carbohydrates and low in fiber.

**Confinement operation**   Operating systems where animals are confined in buildings or cages.

**Conifer**   An evergreen tree that produces cones.

**Consumer** A person who buys goods or services from other people.

**Controlled environment** An environment for animals that is kept at the correct temperature and other conditions to maximize animal comfort.

**Cotton gin** A machine used to separate the cotton seed from the lint.

**Cow** A female bovine that has had a calf.

**Crop residue** Stalks and other debris left in the field from last year's crop.

**Crossbreed** An animal with parents of different breeds.

**Cross-pollination** Pollination that is the result of gametes from different plants.

**Crustacean** Aquatic animals with a rigid outer covering, jointed appendages, and gills.

**Cultivated plant** Any plant that has been planted, tended, harvested, and improved.

**Curd** A milk product consisting of casein and fat; the coagulated part of milk.

**Cut flower** A flower that has been harvested by cutting it off with the stem attached.

**Cutting** A part of a plant that has been cut off for the purpose of propagating a new plant.

# D

**Dairy** Pertaining to products that are derived from milk. Also, a farm where cows are kept for milk production.

**Dam** The mother of an animal.

**Dam breed** A breed of animal that makes good mothers that are used in crossbreeding programs.

**Debeak** The removal of the beak from a baby chicken.

**Deciduous** Refers to a plant that loses its leaves every year.

**Diet** The amount of food taken in during a 24-hour period. The term can be used in referring to either human food or animal feed.

**Disease** Any departure from the normal health of a plant or animal.

**Disinfectant** A substance that kills pathogens (germs).

**Dissolved oxygen** Oxygen that is dissolved in water and is usable by animals that breath through gills.

**Division** A means of plant propagation that involves cutting crowns, rhizomes, tubers, or roots into sections.

**Docile** Refers to an animal that is gentle in nature.

**Docking** The removal of an animal's tail. This is done to help keep the animal clean or to prevent tail biting.

**Dressed** Refers to an animal that has been slaughtered and the carcass cleaned.

**Drug residue** Any amount of a drug that is left in an animal's body.

# E

**E. coli bacteria** Bacteria that naturally inhabit the human colon.

**E horizon**   A mineral horizon, mainly a residual concentration of sand and silt high in content of resistant minerals as a result of the loss of silicate clay, iron, aluminum, or a combination of these.

**Ecology**   The totality or pattern of the interrelationship of organisms and their environment, and the science that is concerned with that interrelationship. The subject is often categorized into plant, animal, marine, terrestrial, freshwater, desert, tundra, and tropical ecology.

**Ecosystem**   The entire system of life and its environmental and geographical factors that influence all life, including the plants, animals, and the environmental factors.

**Ectothermic**   Refers to an animal that gets its body temperature from its environment.

**Endothermic**   Refers to an animal that generates its own body temperature internally.

**Environment**   The sum total of all the external conditions that may act on an organism.

**Environmental Protection Agency (EPA)**   An agency of the federal government that oversees environmental concerns.

**Enzyme**   A substance that speeds up or stimulates a chemical reaction.

**Erosion**   The wearing away of the soil by water or wind action.

**Exotic plant**   A plant that is out of the ordinary.

**Experiment station**   Any of the USDA research facilities associated with state agricultural universities, where new ways of farming, ranching, or rural living are officially tested.

**Export**   Shipment of commodities to foreign countries.

# F

**Farrow**   The process of giving birth to pigs.

**Farrowing crate**   A cage or crate in which a sow is placed for farrowing. It keeps her from crushing the pigs.

**Feedlot**   A lot where cattle are kept for fattening for market.

**Fermentation**   The processing of food by the use of yeasts, molds, or bacteria.

**Fertilizer**   Any organic or inorganic material added to soil or water to provide plant nutrients and to increase the growth, yield, quantity, or nutritive value of the plants grown therein.

**Fibrous root**   A root system which is composed of many fine, branched roots.

**Fingerling**   A fish from 3 to 6 inches long that is used for stocking pods or lakes.

**Flock**   A group of poultry or sheep.

**Flood plain**   The area along a river or body of water that is covered in water by periodic flooding.

**Floriculture**   The cultivation of plants for their flowers.

**Florist**   A person who deals in arranging and selling flowers.

**Flour**    The fine-ground product obtained in the commercial milling of wheat and other grains which consists essentially of the starch and gluten of the endosperm. It is used for baking bread, pastry, cakes, etc.

**Foliage plant**    A plant that is grown for its attractive foliage.

**Food and Drug Administration (FDA)** An agency of the federal government that regulates food and drugs that are sold to consumers.

**Food chain**    The transfer of food energy from the initial source in plants through a series of organisms by repeated eating and being eaten.

**Forestry**    The science of growing trees.

**Free range**    Allowing animals to live in an area unrestricted by a fence.

**Fresh water**    Water that contains no or very little salt content.

**Fruit**    The matured ovary of a flower and its contents including any external part which is an integral portion of it.

**Fry**    Small, newly hatched fish.

**Fumigate**    The destruction of pathogens by the use of liquids or solids that release a vapor.

**Fungi**    Plantlike organisms that have no chlorophyll; they get their nourishment from living or decaying organic matter. Plural of fungus.

# G

**Gaited**    The definite rhythmic movement of a horse such as trot, canter, pace, etc.

**Game bird**    A bird that is hunted for food.

**Gene**    The simplest unit of inheritance. It is composed of DNA.

**Genetic engineering**    The altering of the genetic components of an organism by human intervention.

**Geology**    The study of the Earth, earthy materials, their history, processes, and products. This includes water, air, minerals, and rocks.

**Germination**    The emerging of a plant from a seed.

**Glacier**    A slow-moving mass of ice.

**Greenhouse**    Any of several different types of heated, glass- or plastic-covered structures used for the growing of plants.

**Groundwater**    Water within the ground that supplies wells and springs.

**Growing season**    The number of days from the average date of the last freeze in the spring to the average date of the first freeze in the fall; the period in which most plants grow.

# H

**Hardwood**    Wood from a deciduous tree.

**Heavy soil**    Generally, a clayey soil in contrast with a sandy one. The reference is to resistance and need for greater power in pulling a plow rather than to actual weight or specific gravity which is less for dry clay than for dry sand.

**Heifer**    A female bovine that has not had a calf.

**Herbicide** A substance that is used to kill plants.

**Heterosis** The degree of superiority of a crossbred animal over what would be expected of its purebred parents. Also known as hybrid vigor.

**Hive** A box used to hive a swarm of bees.

**Homogenized milk** Milk that has been processed to blend the butter fat.

**Hormone** A chemical substance produced by an organism that causes a specific effect on the animal or a portion of its systems.

**Horticulture** The branch of agriculture that deals with the growing of fruits, flowers, vegetables, or ornamental plants.

**House plant** An ornamental plant that is grown for its beauty and kept in the house.

**Humus** Organic matter in the soil.

**Hunters and gatherers** People who obtain their food by gathering wild plants and animals.

**Hutch** A cage and shelter to house rabbits.

**Hybrid vigor** *See* heterosis.

**Hydrological cycle** The cycle of water evaporating and condensing and beginning the cycle over again.

# I

**Import** Goods that are brought into this country from another country.

**Incubator** A device that controls the temperature and relative humidity for the hatching of eggs.

**Infection** The invasion of an organism's tissues by a disease-causing agent.

**Inorganic soil** A mineral soil; a soil in which the solid matter is dominantly rock minerals in contrast to organic soils, such as peats and mucks.

**Insecticide** A substance used to kill insects.

**Integrated pest management** A method of controlling pests that uses a variety of means.

**Interiorscaping** The use of plants inside a building to produce a pleasing effect.

**Internal combustion engine** An engine in which both the heat energy and the ensuing mechanical energy are produced inside the engine proper.

**Irrigation** The artificial application of water to soil for the purpose of increasing plant production.

# K

**Koi** Fish of the carp family that are raised because of their colorful appearance.

# L

**Lagoon** A pond used for disposing of animal manure.

**Land-grant university** Any of the state colleges and universities started from federal government grants of land to each state to encourage further practical education in agriculture, homemaking, and the mechanical arts. The mission of these universities is to conduct programs in teaching, service, and research.

**Landscape architect**   A person trained in the art and science of arranging land and objects on it for human enjoyment and use.

**Landscaping**   To beautify the terrain with the planting of trees, shrubs, and plants.

**Landscaping cloth**   A cloth used to prevent weeds from growing among landscape plants.

**Larvae**   The immature insect as it hatches from the egg.

**Layer**   A hen that is raised for producing eggs.

**Legume**   A family of plants that with the aid of symbiotic bacteria convert nitrogen from the air to build up nitrogen in the soil.

**Life cycle**   The changes in life form that an organism undergoes.

**Litter**   A group of young pigs that were born from the same mother. Also the absorbent material spread on the floor of an animal's pen or house.

# M

**Mastitis**   A disease of the mammary glands that is usually caused by an injury.

**Media**   Soil or soil-like material in which plants grow.

**Meteorology**   The study of the physical processes which occur in the atmosphere and of the related processes of the lithosphere and hydrosphere.

**Microorganism**   An organism so small that it cannot be seen clearly without the use of a microscope; a microscopic or submicroscopic organism.

**Mineral**   A chemical compound or element of inorganic origin.

**Mule**   An animal produced by the mating of a jackass and a mare.

**Mutton**   The meat of a mature sheep.

# N

**Natural resource**   Any resource that grows or occurs naturally.

**Navigable waterway**   A stream or body of water that is large enough to allow the passage of boats or ships.

**Nectar**   The sweet substance from flowers. Bees make honey from this material.

**Nematode**   Microscopic, wormlike, transparent organism that can attack plant roots or stems to cause stunted or unhealthy growth.

**Node**   The place on a stem which normally bears a leaf or whorl of leaves.

**Nodule**   A root formation on certain leguminous plants produced by invasion of symbiotic, nitrogen-fixing bacteria. The bacteria furnish the plant with fixed nitrogen compounds and receive nutrient plant juices like carbohydrates.

**Nonpoint source** Pollution that occurs from several sources rather than a single source.

**Nonselective herbicide** A substance that kills all plants it contacts.

**Nutrient** An element or compound in a soil which is essential for the growth of a plant.

# O

**O horizon** Organic horizons above mineral soil.

**Orchard** A field used for growing fruit trees.

**Organic matter** Matter found in or produced by living animals and plants which contains carbon, hydrogen, oxygen, and often nitrogen and sulfur.

**Organic soil** A soil having a high content of organic soil materials.

**Ornamental plant** Any plant produced for its beauty.

**Ovary** The female organ that produces eggs and female hormones.

# P

**Parasite** An organism that lives off of or at the expense of another organism.

**Pasteurization** The process of killing microbes in milk by the use of heat.

**Pasture** An area of grass maintained for animal grazing.

**Pesticide** Any substance used to kill pests.

**pH** A numerical measure of acidity or hydrogen ion activity of a substance such as food or soil. The neutral point is pH 7.0. All pH values below 7.0 are acid and all above 7.0 are alkaline.

**Pharmaceutical** Any substance used to enhance the health of humans or animals.

**Phloem** The tissue of a plant that transports food.

**Photoperiodic** Plants that react to specific periods of light.

**Photosynthesis** The process in plants that converts light energy to chemical energy.

**Pistil** The female element of a flower.

**Point source** Pollution that occurs from a single source.

**Pollination** The transfer of pollen from the anther to the stigma of a flower.

**Pollution** Substances in a body of water, air, soil, etc. that impair the usefulness of it.

**Pork** The meat from pigs.

**Potted plant** A plant that lives its entire life in a pot.

**Poultry** Domestic fowls that are raised for meat, eggs, or feathers.

**Predator** An animal that eats other animals.

**Propagation** Increasing the number of plants by causing them to reproduce.

**Protozoa** A group of one-celled organisms that generally do not contain chlorophyll; sometimes classified as one-celled animals.

**Pruning**   The removing of limbs or other plant parts to shape the plant.

**Pupa**   The stage of an insect's life between the larva and the adult stage.

**Purebred**   A animal belonging to one of the recognized breeds. The animal's ancestry are all of that breed.

# R

**Ration**   All the feed an animal receives during a 24-hour period.

**Refrigeration**   Artificial cooling, either by the application of ice or by utilizing the principle of the latent heat of evaporation.

**Registration**   The process of certifying that an animal is a purebred of a particular breed.

**Renewable resource**   A resource that can be replaced such as trees that are grown as a crop.

**Rennet**   A substance derived from the stomachs of calves and used in the cheese process.

**Resistant**   Designating a plant or animal that can resist disease.

**Rhizobia**   Bacteria that live symbiotically in roots of legumes and fix nitrogen from the air.

**Rhizosphere**   The root zone, the area where microorganisms are most active in increasing the availability of nutrients for plants.

**Roe**   A mass of fish eggs.

**Root cutting**   A cutting taken from a plant root for the purpose of growing a new plant.

**Rooting hormone**   A substance applied to a plant or root to stimulate growth.

**Roughage**   A feed that is high in fiber content and low in carbohydrate content.

# S

**Sand**   Any of the small, loose, granular fragments which are the remains of disintegrated rocks. It may contain a great variety of minerals and rocks, but the most common mineral is quartz.

**Sea island cotton**   A woody plant grown as an annual for its long staple lint cotton used in fine cotton fabrics. Native to tropical America.

**Seed**   The embryo of a plant.

**Seedling**   A plant that has newly emerged from the seed.

**Selective breeding**   Choosing the most desirable plants or animals for breeding in order to develop superior plants or animals.

**Selective herbicide**   A substance that will kill some plants and have little or no effect on other plants.

**Self-pollinating**   A plant that produces and uses its own pollen for reproduction.

**Semen**   The fluid from a male animal that contains the sperm.

**Sexual reproduction**   Reproduction through the means of the uniting of gametes.

**Shade cloth** A cloth used to cover plants to reduce the amount of sunlight.

**Shrub** A plant that has several woody stems and is low growing.

**Silage** A feed made from chopped and fermented corn.

**Silt** Small, mineral soil particles ranging in diameter from 0.05 to 0.002 mm in the USDA system or 0.02 to 0.002 mm in the International system.

**Sire** The father of an animal.

**Sire breed** Breeds of animals that are used primarily as sires in a crossbreeding system.

**Skim milk** Milk that has had almost all of the milkfat removed.

**Smith-Hughes Act** A United States federal law passed by Congress in 1917 which gives federal aid for vocational education in the secondary schools.

**Softwood** Wood from a conifer tree.

**Soil** The mineral and organic surface of the earth capable of supporting upland plants.

**Soil texture** The coarseness or fineness of soil.

**Spawn** The process of a fish laying eggs.

**Squab** The meat from a pigeon.

**Stamen** The male part of a flower that bears the pollen.

**Stigma** The female part of the flower that receives the pollen.

**Stocker** A beef animal that weighs around 700 to 800 pounds that needs additional growth before entering the feedlot.

**Strip cup** A cup used to catch the first milk milked from a cow.

**Succession** The replacement of one type of plants for another in the natural progression of a forested area.

**Surface water** Water that lies on the surface of the ground.

**Symbiotic relationship** A relationship between two different types of organisms that is beneficial to both of them.

# T

**Tap root** The main descending root of a plant from which branch roots grow.

**Teat** The part of the female mammary system from which the young receives milk.

**Terminal tip** The end of a branch, stem, or root where growth takes place.

**Tissue culture** The process of growing a new plant from plant cells.

**Tolerant** The ability of a plant or animal to resist a pest or disease.

**Topiary** The shaping of plants into shapes that are not natural such as into the shape of animals.

**Trade balance** The balance of imports and exports of a country.

**Tree farm** A farm that grows trees for commercial use such as pulp, lumber, Christmas trees, etc.

# U

**Udder**   The exterior portion of a female's mammary system.

**Upland cotton**   A tropical, woody herb grown as an annual for its fiber, the source of much of the commercial cotton in the United States.

# V

**Veal**   Meat from a calf that is less than 3 months old.

**Vegetable**   The edible part of an herbaceous plant.

**Vocational Agriculture**   Generally refers to the curriculum or program in agricultural education designed to offer students at the secondary level the opportunity to explore and prepare for agricultural occupations. The name of the program has now been changed to Agricultural Education. Also, post-secondary and adult programs are recognized as legal components of Agricultural Education.

**Volcanic ash**   The fine particles from volcanic eruptions which, in thick deposits, are the parent material of soil.

# W

**Wean**   To take a young animal away from its mother when it is mature enough to eat on its own.

**Weed**   Any plant that is growing where it is not wanted.

**Wetlands**   An area that stays wet for most of the year.

**Whey**   The watery portion of milk that remains after the curd and cream have been removed.

**Whorled**   Three or more leaves, twigs, etc. that are arranged in a circle at one point on a plant.

**Wilderness area**   An area set aside by the government to preserve the wild state of the area. Access is limited.

**Wildlife habitat**   The environment in which wildlife lives.

**Windbreaks**   Trees that are planted to block the wind from homes, roads, crops, etc.

# X

**Xylem**   Plant tissue that transports water.

# Y

**Yogurt**   A semi-solid fermented milk food product.

# INDEX